ethuen's Monographs on Biological Subjects

eneral Editors : G. R. de Beer, M.A., D.Sc., F.R.S., and
Michael Abercrombie

THE ECOLOGY OF ANIMALS

MONOGRAPHS ON BIOLOGICAL SUBJECTS

General Editor : G. R. DE BEER, M.A., D.Sc., F.R.S.

Reader in Embryology in the University of London

F'cap 8vo.   Illustrated.   Each 4s. net

SOCIAL BEHAVIOUR IN INSECTS. By A. D. IMMS, D.Sc., F.R.S., *Reader in Entomology, University of Cambridge.*

MENDELISM AND EVOLUTION. By E. B. FORD, M.A., B.Sc. *Demonstrator in Zoology, University of Oxford.* (5s. net.)

RESPIRATION IN PLANTS. By W. STILES, M.A., Sc.D., F.R.S., *Professor of Botany,* and W. LEACH, D.Sc., *Professor of Botany in the University of Manitoba.*

SEX DETERMINATION. By F. A. E. CREW, M.D., D.Sc., *Professor of Animal Genetics, University of Edinburgh.* (5s. net.)

THE SENSES OF INSECTS. By H. ELTRINGHAM, M.A., D.Sc., F.R.S., *President of the Entomological Society.*

PLANT ECOLOGY. By W. LEACH. (4s. 6d. net.)

CYTOLOGICAL TECHNIQUE. By J. R. BAKER, M.A., D.PHIL., *Lecturer in Zoology, University of Oxford.* (6s. net.)

MIMICRY AND ITS GENETIC ASPECT. By G. D. HALE CARPENTER, D.M., *Hope Professor of Zoology,* and E. B. FORD.

THE ECOLOGY OF ANIMALS. By CHARLES ELTON, M.A., *Reader in Animal Ecology, University of Oxford.*

CELLULAR RESPIRATION. By N. U. MELDRUM, M.A., Ph.D.

PLANT CHIMAERAS AND GRAFT HYBRIDS. By W. NEILSON JONES, M.A., *Hildred Carlile Professor of Botany, University of London.*

TISSUE CULTURE. By E. N. WILLMER, M.A., *Lecturer in Physiology, University of Cambridge.* (5s. net.)

INSECT PHYSIOLOGY. By V. B. WIGGLESWORTH, M.A., M.D., F.R.S., *Reader in Entomology, University of London.*

PLANT VIRUSES. By KENNETH M. SMITH, Sc.D., Ph.D.

NEMATODES PARASITIC IN ANIMALS. By G. LAPAGE, D.Sc., *Institute of Animal Pathology, Cambridge.* (5s. net.)

THE CHROMOSOMES. By M. J. D. WHITE, M.Sc., *Lecturer in Zoology and Comparative Anatomy, University of London.* (5s. net.)

THE MEASUREMENT OF LINKAGE IN HEREDITY. By K. MATHER, B.Sc., *Lecturer in Eugenics, University of London.* (5s. net.)

PALAEOZOIC FISHES. By J. A. MOY-THOMAS, M.A., *Lecturer in Zoology, University of Oxford.* (5s. 6d. net.)

THE METABOLISM OF FAT. By I. SMEDLEY-MACLEAN, M.A., D.Sc., F.I.C., *Lister Institute of Preventive Medicine.* (5s. net.)

THE ECOLOGY OF ANIMALS

BY

CHARLES ELTON, M.A.

DIRECTOR, BUREAU OF ANIMAL POPULATION, DEPARTMENT OF
ZOOLOGY AND COMPARATIVE ANATOMY,
OXFORD UNIVERSITY

LONDON · METHUEN & CO. LTD.
NEW YORK · JOHN WILEY & SONS INC.

First Published . . . . November 16th, 1933
Second Edition . . . . February 1946
Third Edition 1950

CATALOGUE NO. 4102/U

PRINTED IN GREAT BRITAIN

DEDICATED TO

DONALD BATEMAN

CONTENTS

THE ECOLOGY OF ANIMALS

THE ECOLOGY OF ANIMALS

CHAPTER I

THE SCOPE OF ANIMAL ECOLOGY

ECOLOGY, in the sense of a knowledge of natural history, dates from the earliest times when man began to put two and two together and exploit the natural resources of his surroundings in order to increase material comfort and security. Primitive races still have this knowledge developed to a high degree. The Arawak of the South American equatorial forest knows where to find every kind of animal and catch it, and also the names of the trees and the uses to which they can be put (Hingston, 1932). The Masai of Central Africa knew for hundreds of years that malaria was caused by the bite of a mosquito and that red water in cattle and heart water in sheep were carried by ticks (Percival, 1924). The Eskimo goes on the assumption that diseases in his sledge dogs can be caught from the wild Arctic fox (Elton, 1931b), and knows what time of year to expect the arrival of seals and ptarmigan. The natives of Shansi knew the connexion between periodic irruptions of the sandgrouse and climatic changes and famine and have a saying ' when the sandgrouse fly by, wives will be for sale ' (Rockhill, 1894). This intense understanding of wild life can be matched among some of those men in our own country who spend much of their lives in the woods and fields, and have become familiar with the habits of wild game and birds and fishes. More highly educated people also have for the last two hundred years taken an increasing interest and pleasure in wild life. They study it for its beauty, or to make collections, or because of the fascination of its complexity. Natural history of this sort has a long and often brilliant record and has created a widespread organization of societies and clubs, with correspondingly large numbers of publications.

Ecology represents partly the application of scientific method in natural history, and partly something more. All the knowledge that has been referred to above is folk-lore or practical knowledge or natural history. When the species of animals are properly identified by an agreed system of names, specimens stored in museums for reference, the habitats of the animals accurately described, the information tested by various different workers, ingenious experiments devised to find out causes, then we can say that natural history has begun to be animal ecology. Ecology means the relation of animals and plants to their surroundings. As a science it may be said to depend on three methods of approach : field observations, adequate systematic technique for determining the names of the animals, and experimental work both in the field and in the laboratory. Or we may consider the points of contact that animal ecology makes with other subjects : animal behaviour leading to comparative psychology, the action of physical limiting factors requiring a knowledge of physiology, the description of habitats with its reliance on physical and chemical techniques, and on vegetation studies, and so on. But when all this has been said there remains a growing body of principles which concern the more purely biological aspects of animal life : the inter-relations of animals, numbers, social organization, migration, food, and many others.

These principles form the purely ecological side of biology. In the present book the relations of animals are treated under the successive phases of complication through which investigations on animal ecology usually pass : Surveys (the description of animal communities in different habitats), Animal Inter-relations (the organization of animal communities), Habitats (the vegetation, physical conditions and limits of animal life, including the methods of measuring them), Numbers (including both the collection of statistics, and the analysis of the dynamics of animal populations), and finally the relation of these principles to some practical problems facing the human race. It is usually with problems of numbers that economic biologists have to deal when they are studying animals, and this represents the final and most complex and hitherto least fully understood branch of animal ecology. Although the subject is developing as a pure science, it is not possible to omit a consideration of economic problems. An important part of the ' sinews of war ' for ecological research will continue to come from economic sources, also many of the most striking facts, especially those requiring large-scale organization for their study, e.g. international locust research or marine fishery studies.

Animal ecology began as a science by following rather closely the lines laid down by the earlier work of plant ecologists. Warming's pioneer book on the ecology of plant associations was published in 1896. The first textbook on animal ecology by Adams appeared in 1913, while the first comprehensive book on any animal communities was published by Shelford in 1913. It became clear later on that animal ecology would have to introduce a number of ideas hardly required by botanists. For instance, animals move about, eat each other, display unexpected reactions, and court each other. It came to be realized that the importance of plant ecology to animal ecology lies in its definition of the habitats in which animals live. In a perfectly true sense, animals and plants in nature are bound together into a complex series of biotic communities, whose interdependence is well illustrated by the action of earthworms on the soil and of the acidity of the soil upon earthworms, or by the interaction of nulliporous algae and of corals on a tropical coral reef. This concept is ably discussed by Phillips (1931). But in most investigations it is found advantageous to separate plant and animal ecology in practice. At the same time, owing to the great extent of ecological field problems, team work by plant and animal ecologists is capable of producing excellent joint studies of animal and plant communities. The difference in methods is soon realized by any one who alternately attempts to count the trees in a wood and the wood-mice that live under the roots of the trees.

The wide range of subjects that is called in to assist the completion of any investigation on animal ecology has already been mentioned. Any comprehensive survey has to refer to maps of topography, geology, soil, climate ; agricultural influences such as grazing ; conservation and protection generally ; the vegetation ; and the systematic background upon which such surveys depend for naming species. The wide range of economic problems that are dependent upon ecological studies is illustrated by the following examples taken at random : the question of over-fishing in the Baltic and North Sea, the feeding grounds and reserve population of Antarctic whales, the acclimatization of foreign fish in East Africa and New Zealand, the breeding grounds of the migratory locust in North Africa or elsewhere, the control of tsetse flies that carry sleeping sickness in Africa, the zootropic (host-selection) habits of malaria-carrying mosquitoes in Europe, the climatic (temperature and humidity) limits of the rat fleas that carry bubonic plague or of the larvae of hookworms, the fluctuations of fur supplies in the Canadian forests and in Arctic regions, periodic plagues of cockchafers in Europe, conservation of game animals in North America and in Africa, the biological control (by means of insects) of introduced plants such as prickly pear and St. John's wort in Australia, of gorse and bramble in New Zealand, and of Lantana in Hawaii, the effects of various factors on the increase of blowflies attacking sheep in Australia, the legions of agricultural and forestry pests such as the codling moth, the pear slug, the cotton boll weevil, the gypsy moth, and many more.

Although these practical problems undoubtedly provide one of the strongest and most urgent reasons for developing principles in animal ecology, there are also important points of contact with general studies on evolution and species formation, with psychological theories, while at the same time the exact study of wild animal communities offers in itself a very adequate intellectual reward to the investigator.

The central organization for the study both of plant and animal ecology in the British Isles is the British Ecological Society, which publishes the *Journal of Ecology* and also a more specialized *Journal of Animal Ecology*. In the United States there is the Ecological Society of America which publishes *Ecology*, and *Ecological Monographs*, the latter for larger studies. In Russia a journal was started, but its future is uncertain. On the continent of Europe papers are mainly published in general biological journals, but there are several important publications devoted to aquatic biology among which may be mentioned the *Archiv für Hydrobiologie*, the *Revue Internationale für Gesamten Hydrobiologie und Hydrographie*, and among marine ecological publications the reports of the *Conseil International pour l'Exploration de la Mer*.

The problem of keeping pace with the literature of animal ecology is a serious one, since much work is published in unexpected places. Most of the British work is noticed and abstracted briefly in the *Journal of Animal Ecology*. On the economic side, attention should be drawn to the abstracts contained in the *Review of Applied Entomology*, which covers in its two series Agricultural Entomology on the one hand and Medical and Veterinary Entomology on the other hand. Overlapping somewhat with the latter is the *Tropical Diseases Bulletin*, which contains also very comprehensive abstracts. Some of the literature concerned with freshwater biology in its economic applications is abstracted in the *Summary of Current Literature on Water Pollution Research* issued by the Government.

General books on animal ecology are still few in numbers. The following may be mentioned : Shelford (1913 and 1929), Chapman (1931), and Elton (1927). Shelford's later book mainly concerns the methods of analysing the physical environment and the ways in which it affects animals. Chapman covers the theories of animal ecology, methods, and literature. The book by Elton treats animal interrelations and numbers more fully than the others, while containing less on the physical side, and upon insects.

CHAPTER II

ECOLOGICAL SURVEYS

ECOLOGICAL research sets out to define the relations between animals and their surroundings. What are these surroundings? They include not only the climate and soil and vegetation, but also the other species of animals that live there. The habitat of every animal is partly dead and partly living, and environmental factors are therefore usually classified by ecologists into edaphic and climatic on the one hand, and biotic on the other. In practice the habitat can best be studied along three lines: the physical and chemical factors, the vegetation, and the animal community. It is not possible to obtain a real picture of the life of any species unless we bring into this picture all the other species of animals associated with it. The nature of these associations— involving food, enemies, parasites, competitors, etc. —is discussed in Chapter III. Before such problems can be discussed at all it is necessary to carry out preliminary ecological surveys which enable the animal community of each habitat to be described and defined. When we know that a species lives in certain conditions of climate, on a certain soil, in such and such a woodland type, in company with several hundred other species whose names and habits of life are known, and is itself inhabited by a definite series of parasites, then the stage is set for a discussion of the relation of the species to its surroundings.

2

This complete knowledge has not yet been obtained for any species. One reason is that comparatively few ecological surveys of animal communities have yet been completed. This phase of animal ecology stands to-day at the point where plant ecology stood in 1905, and where geology stood at the end of the eighteenth century. It follows from this that we still know comparatively little about the manner in which animal communities are organized. Certain principles are, however, beginning to emerge clearly, and will be noted later. But in this phase of animal ecology much work remains to be done, work which has a number of attractions for the student, although beset with considerable difficulties. The attractions are that ecological surveys bring the student into direct contact with live animals in a state of nature, while at the same time almost always leading on to special problems whose existence would not otherwise have been suspected. And such surveys afford an insight, that no other method gives, into the astounding richness and complexity of animal life. They also provide a sense of proportion which prevents undue importance being given to one special group of animals.

The difficulties of such surveys lie chiefly in the actual labour involved in collecting and determining the species of animals captured. Such difficulties are partly those of organization, of getting co-operation from experts, and of ensuring that enough field work is accomplished to give a fair picture of the animal community studied ; but there are also technical obstacles connected with proper preservation of specimens, etc., which can only be overcome by experience. For these reasons beginners are strongly advised to select as simple and limited an animal community as possible, rather than to attempt single-handed a survey of some very complex community such as a wood or a lake. Ecological surveys in high Arctic regions have proved the advantages of studying animal communities containing comparatively few species (Summerhayes and Elton, 1923 and 1928). Where a more complex survey is contemplated, it is essential to work in a team or group, which can divide up some of the different sections of the work. Thus, in surveying the animal community of any British vegetation type such as a heather moor or a wood or a swamp, it is usually found convenient to have about three or four different collectors at least. One man can do the birds, another the other vertebrates, a third the insects, and if possible another man the remaining invertebrates. In addition the help of a botanist will be needed, to give a check on the vegetation. This is especially important at the start, in choosing a suitable area for the field work. The vegetation types give a fair indication of many of the other features of the environment, so that it is an advantage to choose vegetation that is representative of some widespread type. By doing this the survey of animals will have more than a purely local significance, and can be compared with later work in other places. It is of vital importance to define very exactly the nature of the environmental factors of the habitats. This subject will be treated more fully in Chapter IV. As regards general methods of surveying, reference may be made to Elton (1927), who gives also a list of systematic books and papers dealing with various groups of British animals. These books and papers mostly contain notes of the technique of studying each group, as well as keys for their identification. It is seldom safe to identify species without expert confirmation. And in all ecological surveys it is essential to retain full collections of specimens for permanent reference. This means in practice that the collections should go to some museum, where they can be safe against casual accident or faulty preservation. This method of depositing the specimens from ecological surveys has been successfully adopted by many expeditions, and should also be more often adhered to by private individuals who carry out ecological work. Thus, the full collections of the Great Barrier Reef Expedition 1928–29 have been placed on permanent record, with their ecological labels, in the British Natural History Museum. It is a considerable advantage also to deposit with such collections copies of the complete notes concerning the survey, since these can seldom be published in full.

In the British Isles the first considerable ecological surveys were undertaken by the Scottish Bathymetrical Survey in 1895–1904, during which James Murray (1910) collected as a side-line the freshwater animals from various Scottish lochs. This work has also been summarized and discussed by Scourfield (1908). Later freshwater surveys were undertaken by other workers on Rostherne Mere in Cheshire (vertebrates by Tattersall and Coward, 1914); in the English Lake District (Pratt, 1898; Gurney, 1923); in Lough Neagh (Dakin and Latarche, 1913) and Lough Derg and also the River Shannon in Ireland (Southern and Gardiner, 1926); in Yorkshire rivers (Percival and Whitehead, 1929 and 1930); and in Welsh streams and rivers (Carpenter, 1927). As part of studies connected with water pollution, surveys have been done in the River Lark in East Anglia (Butcher and others, 1930) and in Welsh streams (Carpenter, 1924 and 1925), and elsewhere. The estuarine surveys to be mentioned overlap into freshwater in several instances. These estuarine surveys include the Tamar and Lynher in Devonshire (Percival, 1929), the Tyne plankton (Jorgensen, 1928), the Mersey (Fraser, 1932), and considerable studies of the Tay and the Tees undertaken by investigators working for the Water Pollution Research Board. In regard to the last, a summary of the animal communities of the Tay estuary has already been published (Alexander, 1932). Walton (1913) surveyed the intertidal animals on the shores of Cardigan Bay in Wales.

Considerable ecological surveys of marine animal communities round the British coasts have been carried out for the Ministry of Agriculture and Fisheries and the Fishery Board for Scotland. Of these may be mentioned the analysis of the bottom fauna of the North Sea by Davis (1923 and 1925) and by Stephen (1923 and 1933) and of the North Sea plankton in relation to herring food by Hardy (1924) and by Savage (1926). Ford and Baker surveyed the sea bottom communities near Plymouth (Ford, 1923).

On land the most comprehensive ecological surveys in Britain have been done for agricultural research purposes. The importance of the plant and animal organisms in relation to soil fertility has been common knowledge since the publication of Darwin's work on earthworms and soil, and the discovery that nitrifying bacteria are devoured by soil protozoa. Surveys of the animals living in soil of agricultural land (pasture or arable or both) have been carried out at Aberystwyth (Thompson, 1924 ; Edwards, 1929), in Cheshire (Cameron, 1913 ; Morris, 1920 ; and Buckle, 1921), and at Rothamsted (Morris, 1922 and 1927). Most of these studies have been quantitative and such soil faunal surveys stand at present as the most exact surveys that exist for any terrestrial animal communities. With them have been combined also certain surveys of animal life in the ground vegetation of pasture land (Cameron, 1917 ; Morris, 1920). Cameron also included in his survey certain other habitats such as meadow and pasture. The only survey of heathland and woodland that approaches completeness for some groups of animals is that by Richards (1926, also in Summerhayes and others, 1924) for Oxshott Common. In this as also in Hardy's plankton survey of the North Sea a valuable attempt was made to work out the inter-relations of animals as well. The Oxshott survey covered also some other habitats besides those men-

tioned, among them bare sand, *Molinia* grass association, and certain aquatic communities. Artificial
habitats have also been surveyed ecologically ;
examples are the survey of the pests of stored cocoa,
dried fruits, and spices in London warehouses
(Richards and Herford, 1930), of sewage (Peters,
1930), and of waterworks (Kirkpatrick, 1917).

It should be understood that we have up to now
been referring almost entirely to complete surveys of
all groups of animals in major habitats, or at any
rate surveys that have aimed at completeness. A
number of workers have confined their attention to
more limited habitats. Examples are the surveys
of apple bark (Light, 1926) and of potato plants
(Walton, 1925), or surveys which have concentrated
on the ecology of one special group of animals, such
as Scourfield's survey of the Entomostracan Crustacea
of Epping Forest (1898), Boycott's survey of the
distribution of some British molluscs (1921), Bristowe's extensive studies of the spider fauna of small
islands round the British coasts (1931, etc.), and
many more.

It would be impossible to mention all the similar
surveys that have been carried out in other countries.
Attention is drawn to a few examples which illustrate
modern trends in methods of survey, and preference
has been given to papers in the English language.

In North America much attention has been given
to the surveying of vegetation and wild life on the
Rocky Mountains, where the zones of vegetation are
called ' life zones '. Each of these zones contains
within itself a multiplicity of smaller divisions of
habitats. Thus the ' Canadian ' zone contains both
conifer forest, grassland, marsh, and lakes. A good
many such studies have appeared in *The American Fauna*, the serial publication of the United
States Bureau of Biological Survey. A complete
survey of the vertebrates of the Sierra Nevada
mountain life zones has been done by Grinnell and

Storer (1924). Other ecological surveys that may be mentioned are those of Zion Canyon in Utah (Woodbury, 1933), of cave animals in the Carlsbad Cavern in New Mexico by Bailey (1928), of the vertebrates of the Kara Kum Desert in Turkestan by Kashkarov and Kurbatov (1930), of prairie and poplar wood animals in southern Manitoba by Bird (1930), of sand-dune arthropods in Finland by Krogerus (1932), of coral reef and mangrove swamp animals on the Low Isles in the Pacific by Stephenson and others (1931), of marine and littoral communities in Massachusetts by Allee (1923), and of Arctic heath and willow scrub by Longstaff and others (1932).

We have now glanced at the position of animal ecology surveys in the British Isles and abroad. It can be seen that rapid progress is being made in what is a comparatively fresh field. We may pause a moment to inquire what general impression is left by a perusal of all these surveys of different animal communities. First we notice that very few workers, even when they wished to, have been able to make complete surveys of all animal groups living in one general habitat, let alone those living in one area.

Secondly, although most of the surveys are of high standard, they have been undertaken for a variety of motives, and to some extent form a random series. Many large gaps in our knowledge remain. Thus in the British Isles, no complete survey of all the animals of an oak wood, of a beech wood, of *Festuca* grassmoor, of hedgerows, of reedswamp, of sand dunes, or of ponds, has yet been carried out. From the point of view of comprehensive ecological surveys of animal life, the British Isles still remain largely virgin country to the scientist. There is no central museum or office in which all these survey operations can be co-ordinated. This field of work offers also the fascination of exploring the animal communities on the hundreds of islands, large and small, which make up the British Isles. On many of these islands, isolated at various intervals since the last Ice Age, specially limited faunas are found ; some of them also containing peculiar island races of animals (e.g. bank voles on Skomer I., mice and wren on St. Kilda, etc.). A third point, already mentioned, is the importance of having reference collections to back the published statements of surveys. This importance cannot be too widely realized. Finally out of this growing mass of survey data, certain general scientific conclusions are beginning to appear. To these we may now turn.

No one working in the country can fail to be impressed by the great importance of vegetation in controlling the conditions under which animals live. We have only to compare the conditions of temperature and humidity and light on a heather moor with those in a pine wood, to realize this. At the same time, as all animals depend ultimately on plants for food, the presence of a particular species of plant tends to fix certain species of animals to it, and indirectly to control the presence of predatory forms and parasites. In a general way, therefore, the distribution of animals is highly dependent upon the distribution of plants. In earlier surveys it was generally assumed that each vegetation type would have a community of animals more or less exclusively attached to it. This general attitude was reinforced by the publication in 1913 of Shelford's pioneer study of the animal communities in the region of Lake Michigan, in which he stressed the importance of physiological reactions of animals. He defined animal communities in different habitats by certain indicator species which were confined to a narrow range of habitat. It has now been realized by most ecologists that the inter-relations of animals are at least as important in determining whether they can live in a certain habitat. A rook feeds mainly on pasture and arable land, and nests and roosts in woods.

There is no suggestion that the climatic conditions in either of these types of habitat greatly influence its choice. In the one instance the rook goes for its food, in the other it goes to a high place where ground enemies cannot reach its young, or itself while asleep. A parasitic Tachinid fly or Chalcid wasp cannot live and breed unless appropriate hosts are there. The herring follows the plankton, and so do whales.

In analysing the lists of animals of different major habitats that result from ecological surveys, we find that a large number of species are not confined to one life zone or vegetation type or to any narrow set of physical or even biological conditions. At the same time each animal community presents characteristic features which enable us to speak of the community of an oak wood or of a sand dune. The characteristic facies that we meet with is given to it by several different features. First there are the animals that are *exclusively confined* to the community, and these often have special adaptations for life in such places. An example is the nutcracker (*Nucifraga*), a bird of northern Europe and Siberia, whose beak is adapted for the sole purpose of breaking the seeds of cedar trees. The Siberian race has a thinner beak than the European race, and this is correlated with the geographical races of cedar, the eastern having a thin shell and the western a thick shell (Formosof, 1933). Secondly, some animals, although occurring elsewhere, are *more abundant* in the community than outside it. Wood-mice (*Apodemus*) are extremely common in woods, though also found outside them, in gardens and hedgerows and occasionally in fields. Finally there is another thing that gives a characteristic appearance to an animal community. The animals stand in a special relation to one another ; that is, the community has a certain *organization* or *structure*. This would be called its economic structure, if we were studying a human community. The second class of animals, those *characteristic* of a community though not confined to it, play a very important part in determining the structure of the whole, since their numbers will influence the numbers of other species that depend on them for food, or are preyed upon by them, or compete with them. We may say, then, that where species have only a narrow range of environmental conditions that they can live in, they tend to become confined to one particular major habitat or minor habitat (e.g. a plant association or one species of plant). But this does not always occur : studies on Rocky Mountain life zones have shown that many species end their altitudinal range irrespective of the limits of the main vegetation zones. Species with a wider range of physiological reactions range over more than one major habitat, and their specialization is shown more by the relative numbers of individuals that they are able to maintain in each. The combination of specialization, varying numbers, and inter-relations of species, gives each animal community its characteristic facies and structure.

A further conclusion concerns the difference between animal and plant associations. The lists of animals collected during a survey are generally swollen by the presence of many accidental visitors from other places. Animals move about a great deal, both passively (as when spiders are blown about on gossamer) or else actively (as with flying beetles and flies or with migrating birds). The result of this is that animal communities can hardly ever be studied as isolated units, owing to the constant action of these outside influences. On the fjaeldmark tundra of high Arctic regions, the food supply of small spiders is greatly increased by Chironomid flies whose larvae are aquatic and whose adults emerge and fly about over land. In summer in England you may see mayflies flying over fields by a river and being snapped up by sand martins which make their nests in the sides of sand and gravel pits higher up on the valley sides. Many insects visit flowers in one habitat but breed in another. In this sense, therefore a study carried out on the animal community of a single major habitat creates for working purposes an arbitrary boundary which does not exist in nature. To a certain extent, but a much lesser one, the same is true of plants, when seed dispersal is taken into consideration.

While realizing that the usual animal community surveyed is partly an artificial segment of a larger animal world selected for intensive study, we may next enquire how many species of animals live together in such a community. This information can be obtained fairly easily from the results of surveys already mentioned. Only it has to be remembered that these are seldom quite complete, and that they work within certain conventions. Thus soil investigators do not usually include the soil protozoa in their lists of animals. These receive separate study. The same is true of ecological surveys of fresh water, where the micro-fauna is tacitly omitted in many instances. In the table given below, the numbers of species collected in various animal communities is listed. The examples have been selected at random except in so far as some surveys do not give the results in a way that can be summarized. No parasites are included.

THE NUMBER OF SPECIES IN AN ANIMAL COMMUNITY

Habitat	Survey	No. of Species
Dry tundra (fjaeldmark), North Spitsbergen, Lat. 79° 50′	Summerhayes and Elton (1928)	29
Arctic heath, West Greenland, Lat. 64° 40′	Longstaff (1932)	82
Arctic willow scrub, same area	Do.	68
Arctic drift-line, seashore, same area	Do.	25
Calluna heath, Oxshott Common, Surrey, Lat. 51° 20′	Richards (1926)	105

Habitat	Survey	No. of Species
Bare sand, same area . . .	Richards (1926)	66
Pine wood, same area . . .	Richards (in Summerhayes and others, 1924)	80
Aspen parkland, mature poplar, Southern Manitoba, Canada, Lat. 50° 25′	Bird (1930)	c. 140
Prairie, same area	Do.	c. 131
Rotten fencing posts, near Oxford	Richards (1930)	56
Willows growing out of fence posts, same place	Do	44
Permanent pasture soil, Cheshire	Morris (1920)	63
Surface herbage on same	Do.	106
Arable soil, Rothamsted, Hertfordshire	Morris (1922) 60 and 72 Morris (1927) 25–31	
Dried fruit, cocoa, spices, in London warehouses	Richards and Herford (1930)	126
River Wharfe, Yorkshire . .	Percival and Whitehead (1930)	136
River Lark, East Anglia . .	Butcher and others (1930)	68
Plateau streams in Wales . .	Carpenter (1927)	84
Spring brooks, coast area, Wales	Do.	122
Large rivers, coast area, Wales .	Do.	57
Intertidal (all zones), shore of Cardigan Bay, Wales	Walton (1913)	156
Mud, sea-bottom, Plymouth. England, Lat. 50° 20′	Baker (in Ford, 1923)	35
Sand, same area	Do.	36
Shell gravel, same area . . .	Do.	30
Sea-bottom, Dogger Bank . .	Davis (1923)	75
Mud, sea-bottom, Woods Hole, Massachusetts, Lat. 41° 50′	Allee (1923)	128
Sand, same area	Do.	139
Gravel, same area	Do.	63
Zostera beds, same area . . .	Do.	138
Rocks and rockweed, same area	Do.	173
Wharf pilings, same area . .	Do.	123
Coral reef flat, Low Isles, Great Barrier Reef, Lat. 16° 23′ S.	Stephenson and others (1931)	79
Mangrove swamp area, same islands	Do.	40

It is not possible to point out the detailed allowances that would have to be made to make these different surveys comparable. But certain facts stand out clearly. First the comparatively low number of species which make up any animal community of a major habitat such as a wood, a heath, a coral reef, or a river Taking the figures at their face value, we observe that the highest recorded number of species is 173, while the most frequent values lie between 60 and 140. This generalization is of value, since it indicates roughly the framework within which an animal community is organized. We may expect to find usually not more than 200 species of animals (taking forms visible without a microscope). In some rich marine habitats, and likewise in woods and even more so in tropical forests, we may expect to find a good many more species than this. Probably the number of species of animals in a community which has well-defined vegetation usually exceeds but does not greatly exceed the number of plant species found there. For example, Walton found 69 species of algae on the coast of Cardigan Bay, and 156 species of animals. The number of plant species found on chalk grassland by Tansley and Adamson (1926) was 151, but only from 24 to 75 occurred at each station studied. The number of animals on grassland almost certainly exceeds a hundred, as shown by the work of Morris and Cameron. But, in the high Arctic, plant species outnumber animals (North Spitsbergen at one station, 52 species of plants, and only 27 species of animals, Summerhayes & Elton, 1928). In tropical rain forest the scales would be heavily weighted the other way.

We seem then to be in a position to generalize as follows from the ecological survey work so far done. The number of species collected by present methods of surveying in an animal community of any major habitat does not usually exceed about 150 in temperate regions. In the high Arctic the number is much less, also on certain very unfavourable habitats further south. But the number of species

on a Greenland heath is about the same as that on
a heath in the south of England.   We conclude, therefore, that except in very unfavourable conditions for
life, the number of different kinds of animals that
can live together in an area of uniform type rapidly
reaches a saturation point.   In habitats of great
complexity, which include important minor habitats
(as with vertical layering in a wood) the saturation
number is much higher.   We do not know how far
this generalization applies to tropical regions.   The
reasons for this saturation point in number of species
must be sought in the study of the structure of animal
communities and the dynamics of animal populations,
and the precise values given to community lists will
also depend on the conventions adopted for excluding
casual and accidental forms.   An analysis of the
faunas of soil as given in various surveys shows that
although the total number of species remains roughly
the same in soil in different parts of England, the
actual species composition varies enormously.   Aberystwyth and Cheshire surveys of pasture soil animals
(85 and 58 species respectively) showed nearly the same
number of species but only eight of them—about a
seventh to a tenth of the faunas—were common to
both areas (Thompson, 1924).   The saturation figure
is apparently not directly dependent on the numbers
of individuals present, for Morris (1922) found that
on unmanured soil at Rothamsted there were 60
species and about 5,000,000 individuals per acre,
while on manured soil there were 72 species and about
15,000,000 individuals per acre.   All this goes to
support the idea that there is some important principle
involved in the stability of the total number of species
in an animal community.   Further surveys are required and also the standardization of methods, before
the problem can be carried much further.   Much
work along these lines has been done in plant ecology,
and it appears that the statistical methods developed
in the study of the species composition and abundance

of plant associations could be usefully used or adapted for animal studies.

It is of some interest to note that ecological surveys of parasites can be carried out on the same principles, though not by the same techniques, as those of free-living animals. For instance, a survey of the parasite fauna of wood-mice (*Apodemus sylvaticus*) in about 450 acres of woodland near Oxford (Elton & others, 1931) disclosed some 47 species of parasites (including one kind of spirochaete but not the other kinds, or bacteria). The composition of this community resembles in principle that of any other animal community in a major habitat such as a wood or a moor or a pond. Some species are confined to *Apodemus* as host; others are especially abundant on *Apodemus*, but also occur on other hosts; while some are accidental or at any rate occasional visitors from species of voles and shrews. The relation of wood-mice to other animals in the community is shown by such parasites as the tapeworm that occurs adult in domestic cats and larval in wood-mice, and by the existence of an overlapping system of flea-contacts between such widely different forms as the mole, the common shrew, the bank vole, and the wood-mouse. Such contacts have, of course, an important influence on the possible spread of micro-parasites during epidemics. Again, the parasites found in one host can be classified according to the different minor habitats they select. The wood-mice had the following distribution of species :

Ectoparasites :
  On ear : 1 tick larva
  On fur : At least 12 mites
       1 tick (adult)
       1 beetle
       11 fleas
       1 louse
  On anus and genital organs : 1 mite
  Under skin of limbs : 1 mite

Endoparasites :
   In liver : 1 tapeworm larva
   In stomach : 1 roundworm
   In small intestine : 3 roundworms
                        2 tapeworms
                        3 flatworms
                        2 flagellates
                        1 ciliate
                        1 amoeba
                        2 coccidians
   In coecum : 1 ciliate
   In kidney : 1 spirochaete
   In blood : 1 trypanosome

It will further be noticed that the same parasite may occupy a different position in the body when larval and when adult. It is interesting to note that two other surveys of mammal parasites have given similar figures, e.g. 50 species of parasites for the suslik or ground squirrel (*Citellus pygmaeus*) of south-east Russia (Sassuchin and Tiflov, 1932) and 27 for English rats (*Rattus norvegicus*) (Balfour and others, 1922).

We now begin to perceive the sort of way in which animal communities can be grouped into various grades of unit : the host-population of one species, the community of a major habitat such as a plant association, the community of the life zone or of the sere (that complex of habitats which is related by ecological succession—see p. 37). The chief principle which is beginning to emerge from this analysis of ecological surveys is that in any fairly limited area only a fraction of the forms that could theoretically do so actually form a community at any one time. The list of, say, soil animals from pasture land all over England may run into thousands, but in one field it will not usually exceed a hundred. ' Many are called, but few are chosen.' The animal community really is an organized community in that it apparently has ' limited membership '. An important question at once arises, as to whether the species composition varies not only spatially but from year to year in one place. We know that abundance of the component species varies greatly, and it is to be expected that as one falls below the level of density which enables it to continue permanently, another species may arrive and take its place. If this is so, an ecological survey based on the results of ten years' collecting might give a somewhat misleading picture of the total number of species ever present in one year. There is a suggestion of this phenomenon in some of the surveys given in the table above, and the question requires careful testing.

CHAPTER III

ANIMAL INTER-RELATIONS

THE completion of a thorough ecological survey of a single fairly uniform habitat leaves us in possession of a list of all the species living there. As was pointed out in the last chapter, these lists are characteristic for each kind of habitat in any one area. That is, each kind of habitat has its distinctive animal community living in it, this distinctiveness depending partly upon the ecological specialization of certain species, and partly on the greater abundance or scarcity of the species. We also saw that the list of species from any one area is a good deal smaller than might be expected, and that the number of species in such a community (not counting the parasites or microscopic forms) often lies between sixty and a hundred and forty. In order that the survey lists may be given a meaning, we have to inquire now how this community is organized. All animal communities are *organized* in a certain way—that is, they have a certain *structure*. This structure is fundamentally similar in very widely different habitats, and it depends chiefly upon the manner in which animals are inter-related ecologically. In analysing the list of animals from a community, in order to understand its structure and organization, several different principles can be used.

First, it is convenient to separate the animals that come out by day from those that come out by _corpuscular - active during day + dusk_

night. In an oak wood, for instance, there are certain beetles, leafhoppers, small warblers, and the sparrow-hawk that come out by day. At night there are moths, wood-mice, nightjars, owls, and bats. These diurnal and nocturnal communities are to a large extent independent, although there is overlapping of three kinds : some animals are active mainly at dusk—as the noctule bat and one of its foods the dor-beetle, some animals come out both by night and by day—as the grass-mouse or vole, while some night animals prey on day animals and while they are at rest, and vice versa—as the barn owl when it raids sparrow roosts at night and the song thrush when it attacks earthworms by day. In spite of these exceptions the broad distinction remains. Definite nocturnal communities are found in all terrestrial habitats except in polar regions and possibly the soil. In fresh water the division is also of importance. In the sea, tidal rhythms play a dominant part in the lives of shore animals, but it seems probable that these forms also have important day and night rhythms which, interacting with the tidal rhythms, may be responsible for some of the curious monthly or fortnightly periodicities in reproduction that have been recorded in a good many marine species (see Munro Fox, 1932, and Amirthalingam, 1928). In the depths of the ocean or in deep lakes, animals live permanently in the dark, while Arctic animals live in permanent daylight during the summer months. The whole problem of these short rhythms of activity in animals has as yet received comparatively little systematic attention. A careful study of certain nocturnal insects and other invertebrates in a beech-maple forest in Ohio by Park and others (1931) has stressed the importance of internal activity rhythms in insects that come out at night, while similar studies have been made on deer-mice in America (Johnson, 1926) and wood-mice in England (Elton, and others, 1931) and on fish and other animals in aquariums by Boulenger (1929). Keeble (1910) demonstrated the persistence of tidal rhythms of activity in the marine flatworm (*Convoluta roscoff-ensis*) when brought into the laboratory. The whole question is of great importance in medical ecology. Many malaria-carrying mosquitoes are active only at night or at dusk, while the tsetse flies (*Glossina*), which carry the trypanosome of sleeping sickness in Africa, are active during the day. There is one species of pathogenic *Filaria* (a roundworm) whose larvae abound in the peripheral blood-circulation of infected human beings during the day and are spread by the bite of a diurnal fly, while another species appears in the outer blood at night and is spread by a mosquito. Similarly, the itch-mite of scabies (*Sarcoptes scabiei*) becomes active under the skin at night, causing intense irritation.

It is important, then, to realize that most communities are of a dual nature, having two different sets of animals working as it were in shifts, and coming into activity alternately. The next step is to classify animals by their food-habits. Animals differ biologically from plants in the variety of their food-habits, and the fact that they prey to a large extent on other animals and on plants. It is these facts that make the methods of animal ecology totally different from those of plant ecology. The botanist determines the available food of plants by studying the soil and the sunlight. The animal ecologist has to study the plants themselves and most of the animals as sources of the food of animals. The first type of work can be done by physical and chemical methods. The second involves an enormous amount of natural history observation and experiment, and at the same time makes the study extremely fascinating owing to the number of curious adaptations and habits that are met with. In seeking to understand the general organization of animal communities the student will have to resist the inclination to follow up these fascinating side-tracks. Although the first half-hour of field work will suggest hundreds of intricate problems connected with the details of food-habits and other phenomena, it is necessary for the moment to concentrate on the very general principles that can be seen at work among animal communities in nature.

Animals may be classified according to their food-habits into *herbivores* and *carnivores* and *scavengers*. Carnivores can be roughly divided again into *predators* and *parasites*, but the distinction is not a sharp one, although it is of fundamental importance ecologically, in connexion with problems of numbers in an animal population. It has been pointed out (Elton, 1927) that this distinction between parasites and predators depends on the relative sizes of the carnivore and its prey. If the latter is relatively large it affords, besides a source of food, a possible home and also a means of transport. But there are numerous intermediate types of carnivore which cannot be logically classified either as predators or parasites. Examples are the lamprey, blood-sucking insects, queen ants that become temporary 'social parasites', and many more. An extremely important borderline class is that of insects that parasitize other insects. These are sometimes called *parasitoids*. They live in their host as larval parasites, but when they grow beyond a certain size emerge as free-living insects, having destroyed their host. Scavengers live upon dead or dying animals and plants. Examples are the jackal, the burying beetles, and many Collembola.

Herbivorous animals can be further classified into a huge number of different types. Every species of plant is probably attacked by some kind of animal. The first thing to do in analysing the food habits of animals from an ecological survey list is to determine the forms which are strictly dependent upon one species of plant. Others will be found to occur on several species. Then herbivores can be classified in another way, according to their general method of feeding, or the general type of vegetation that they live upon. There is a very large class of insect plant-suckers, of which the plant-lice or aphids and certain other bugs are the most important. Other groups of insects make mines in the leaves of plants, others again form galls as do also certain species of mites, while certain species live on lichens or on fungi. The list of different types of food-habit among herbivores could be multiplied indefinitely. Enough has been said to indicate the importance of ecological _niches_. The niche means the mode of life, and especially the mode of feeding of an animal. It is used in ecology in the sense that we speak of trades or professions or jobs in a human community. And it is these niches which give any animal community much of its special structure and appearance. It will be noticed that the idea of a niche is a purely ecological one, not taxonomic. Thus pollen feeders which visit flowers may include representatives of many different orders of insects. Grass-eaters in pasture land include cattle, rabbits, voles, grass-hoppers, beetles, moths, snails, etc. Grinnell and Storer (1924) successfully developed the idea of ecological niches during their study of the life zones of the Rocky Mountains. They found that where several species of any one genus of mammal (e.g. ground squirrels) occurred in the same life zone, they usually occupied different ecological niches and did not come into direct competition. On the other hand, the same niche is often occupied by widely differing forms of animal. These forms are in direct competition with one another for the same food. It seems probable that, since the number of niches that exist at all for vegetarian animals in a community is limited, we have a glimpse of one of the reasons why the number of species is rather limited.

When we come to consider the predatory forms we can discern other reasons why the number of different species is limited. Each herbivorous animal is usually preyed on by a good many predators, but the latter do not usually restrict themselves to one host only. Longstaff (1932) found that Passerine birds in West Greenland (such as the Lapland and snow buntings and wheatear) showed a very wide range of choice in their food. Pentelow (1932) has shown the same thing in the case of the brown trout in England. The trout studied by him appeared to eat practically any form of small animal available in the habitat. One of the chief factors limiting the choice of food in these polyphagous forms appears to be size. Thus the trout only ate young crayfish and not older ones. Sticklebacks eat young *Gammarus pulex*, but only up to the second or third moult stage. An Arctic fox will attack a ptarmigan and eat it ; it will attack a pink-footed goose, but only to try and drive it away from the eggs. The Arctic hare it cannot destroy. This sort of relation between the sizes of predators and the animals they prey upon is very important in splitting up the animal community into food-niches. The limitations placed by size of food on feeding together with other special food-preferences give rise to *food-chains*, leading usually from smaller to larger forms, and starting of course from some herbivorous or scavenging form, which in turn depends directly or indirectly on plants. An example of a Sub-arctic food-chain is a springtail (pollen-feeder or scavenger) eaten by a hunting spider which is in turn eaten by a Lapland bunting. On an English heather moor the species would be different but the niches the same, e.g., the bird would be a meadow pipit instead of a Lapland bunting. An example on pine wood country in Surrey or Hampshire is the small fly that is caught by a small spider, the latter by wood-ants, the wood-ants by a green woodpecker, and the latter possibly by a sparrowhawk.

In the North Sea, diatoms (which are algae) in the plankton are eaten by the copepod *Pseudocalanus*, which is eaten by sand-eels, and the latter by the herring, while the herring is captured by man and sea-birds and dogfish, the latter itself being also caught by man.

This phenomenon of food-chains was first pointed out by Shelford (1913), who constructed a theoretical *food-cycle* diagram for some of the animals of some Illinois habitats. It was not until some ten years later that more exact food-cycles began to be worked out for animal communities. In no instance has it been possible to work one out completely, but the method is a useful one, and enables the organization of an animal community to be more clearly seen than would otherwise be possible. Food-chains are seldom simple and self-contained. Usually several preys and several predators interact at each stage. Two generalizations can, however, be made about them. First, there are seldom more than half a dozen stages in the sequence from herbivore to the final predator that has no predatory enemies. Usually there are not as many as this. Thus the snowshoe rabbit in Canada is a vegetarian, and is preyed upon by the lynx and the fox and various birds of prey, none of which have further enemies. When we start with a smaller animal such as a copepod or a springtail or an aphid, there are usually more stages in the complete food-chains which come to depend upon them. The limited number of stages is due primarily to size considerations. There is considerable difference between the sizes of most predators and their preys, and after a food-chain has passed through several stages, size limits begin to be reached. The second point, which further explains the first, is that the numbers of individuals gets smaller as each stage in a food-chain is reached. The reason for this is that every animal has to produce not only enough young to provide against non-biotic checks, but also

Key industry — as the base of the Pyramid of numbers.

the large margin necessary to support this super-structure of predatory animals. At each stage the animals are larger, they breed more slowly, and there is less of the original plant-produced living matter to be used in providing the margin. This phenomenon has been called the *pyramid of numbers* (Elton, 1927), and is an essential feature of the structure of animal communities. It does not of course apply to food-chains of parasites, which get smaller and smaller with individuals more numerous at each stage (parasite, hyperparasite, etc.).

Just as man has been compelled to adopt methods of artificial limitation of numbers in his population, now that wars and famines and infant and adult mortality from disease organisms no longer provide a natural check, so animals which have no natural enemies, or which are comparatively immune from them, tend to adopt systems of limitation of numbers. This forms an interesting subject which cannot be more than mentioned in passing. Reference may be made to the territory systems found among many birds during the breeding season, e.g., in kingfishers, hawks, warblers, etc. (Howard, 1920), among carnivorous animals such as tigers, badgers, and in certain social insects (e.g. wood-ants, Elton, 1932). Territory should be considered as a phenomenon which often leads to limitation of population, but has not necessarily been evolved solely as an adaptation for that purpose.

Much ecological work is being done now upon the exact food-habits of animals, but a great deal more needs to be done before complete preliminary maps of the food-relations of animal communities can be drawn up. It will therefore be realized that few attempts have been as yet made to make such maps, and that these are at best rough and very incomplete. They do, however, enable the conclusions to be drawn which have just been outlined—the ideas of the complex food relations of animals among themselves (as distinct from plants which have in the main a common food supply), food-chains which are built up into complex food-cycles for a whole community, the limited number of stages in each chain, the idea of niches (which applies also to other habits than feeding), the pyramid of numbers in which each species in a progressive food-chain becomes lower in its density of numbers, the existence of habits which partly control population among the species at the end of such chains, the corresponding chains of parasites, and throughout these the dominant importance of the size of animals. As a result of these principles we find that the organization or structure of an animal community is not widely different in almost any habitat which supports a rich fauna at all. Even in intertidal marine communities where sessile animals play a dominant part, food-chains form the most striking feature in the life of the community. Food-cycle maps have so far been published for all or parts of the following communities : temperate forest communities in Illinois (Shelford, 1913), high Arctic fjaeldmark, and associated habitats (including freshwater) in Spitsbergen, North-East Land and Bear Island (Summerhayes and Elton, 1923, 1928), aspen parkland and prairie in Manitoba (Bird, 1930), young pine trees on Oxshott Common in England (Richards, 1926), the vertebrates of the Kara Kum desert (Kashkarov and Kurbatov, 1930), and for the plankton of the North Sea in relation to herring food (Hardy, 1924).

These intricate series of inter-relationships of whose nature we are only now beginning to see a glimpse, have inevitably given rise during evolution to innumerable special adaptations connected with attack and defence. These adaptations have been studied in great detail, and form a subject of their own—especially in connexion with the widely debated but firmly established theory of *mimicry* (see Carpenter and Ford, 1933, in the present series). Mimicry forms one of the special kinds of protective resemblance of one animal by another, in which the chief element is a bluff—the imitation of the strong by the weak or the distasteful by the edible. As a result of such adaptations the actual edible qualities of wild animals do not always become closely adjusted to the potential enemies that exist in the community.

We have not so far referred much to the special inter-relations which exist between members of the same species. The nature of direct competition between individuals comes under population problems in the community and is discussed in Chapters V and VI. Under this heading also comes the subject of animal aggregations, although these have probably formed the basis of sub-social behaviour upon which evolution has worked in the production of social groupings within the species (Allee, 1931). Social inter-relations within the species are of two kinds, of which the first is the sexual relation. The existence of the sex chromosome mechanism of sex-determination usually produces equal numbers of males and females in a species; but from an ecological point of view, as Dr. John R. Baker has pointed out to me, the fertilization of sufficient females to keep the species above a critical density of numbers could be accomplished by a fraction only of these males. This is, in fact, the case with polygamous species such as blackgame and *Gammarus* and others. We know that a single male rabbit can successfully fertilize forty females in one day. The superabundance of males generally found in nature has led to the development of competition between males for possession of the females. We find among insects and birds and mammals and most other groups above the lowest, more or less highly developed courtship ceremonies —or to use a non-committal term, epigamic behaviour. Competition among the males is only one reason for this behaviour, and there are a number of obscure problems connected with it as yet unsolved (e.g. the

generally found coyness in female insects during courtship (Richards, 1927), and special features found in
territorial birds). Such epigamic behaviour plays a
very important part in the lives of wild animals,
takes up considerable parts of their time during the
breeding season and sometimes outside it, and cannot
be left out of consideration from an account of the
organization of animal communities. For instance,
certain insects frequent certain habitats only in order
to carry out epigamic ' ceremonies ', and while there
form the food of other forms (cf. Richards, 1930).
The conditions necessary for the successful carrying
out of such courtship may also determine whether
a species can breed in a habitat, while the field
observation of such behaviour has a peculiar fascination for many naturalists.

The second type of inter-relationship concerns social
groupings and the development of castes. Wheeler
(1928) estimates that genuine social life has developed
independently in at least thirty different groups of
insects. It is most highly developed in the ants and
bees and termites (see Imms, 1931, in the present
series). Social life in ants has tremendously important effects upon the general organization of life in
some animal communities, in relation to destruction
of many species of animals on a large scale, ' farming
habits ' with aphids, coccids, etc., other agricultural
practices, e.g. the growing of fungi for food, and
indirectly in the construction of large habitations in
which many other forms of animals live as guests
or parasites of varying degrees of dependence. Reference should be made to Wheeler (1922, 1928) and
Imms (1931) for fuller accounts of social insects and
to Alverdes (1927) for information both about insects
and about other groups. The chief effect of social
habits on the ecological organization of an animal
community is through the greatly increased power
that it gives one species over many others, with the
result that the social group sometimes becomes as

dominant economically as does man. The spectacle of insects forming complex civilizations not dissimilar to our own in certain features of their economic organization is of the greatest interest.

CHAPTER IV

HABITATS

WE have seen in the last chapter that the habitat of every animal can be divided for convenience into three parts : the animal community, the vegetation, and the physical factors. It is sometimes said that animal ecology is only the study of the physiology of whole organisms instead of the physiology of particular organs or organ systems. But it can easily be seen that this is a very limited view to take of the subject, since such physiological studies, although of great importance, cover only part of the environmental complex that determines the conditions in which an animal can live and breed. Ecology covers such a wide field, and experimental work is usually so difficult and costly and takes so many years to complete, that it is not surprising to find that a large specialized section of ecological research has grown up, which deals only with the effects of physical factors upon animals. It is clear that in applying such data to the explanation of animal distribution and numbers in the field, the vegetation and the mode of organization of animal communities must also be taken into account. We have dealt briefly with animal interrelations, and something has now to be said about vegetation.

In surveying the possible habitats available for animals, we are struck by the dominant influence of vegetation in creating large and comparatively uniform habitats all over the surface of the earth. Vegetation has two main effects (apart from food and shelter to animals). It modifies the natural climatic and soil conditions and to a certain extent smooths out their temporal fluctuations. At the same time the phenomenon of dominance, by which one or two species of plants dominate the rest in the competition for light and food, produces rather sharp boundaries between different plant associations. This produces comparatively sharp differences between the environmental conditions, e.g., of temperature and moisture in each habitat. Through the influence of vegetation most of the earth is divided up into a patchwork of habitats, each comparatively uniform in conditions and each rather abruptly separated from the next. This is usually what we mean when we speak of 'major habitats': the area covered by a particular vegetation type with its characteristic dominant species, and corresponding association of other subordinate species. When we speak of vegetation making conditions uniform, it should not be forgotten that it also creates a variety of minor habitats, partly through the variety of plant species and partly through vertical layering, as in woods. It is the general climatic variations that are toned down.

Another way in which vegetation affects animal habitats is through ecological succession. Ecological succession takes place also independently of vegetation, as when a river erodes its banks and lays down sediment elsewhere, or when a sand dune advances and replaces intertidal areas, or when the lime gets gradually leached out of soil. But these physical changes chiefly have the effect of creating new bare areas on which the development of vegetation takes place in a fairly orderly sequence which is characteristic for any particular climate and soil and geographical region. This sequence of plant associations is called a 'sere'. Through the effects of ecological

succession, which in later stages is usually brought
about by the plants changing their own soil conditions through accumulation of humus, decreased light
killing subordinate species or preventing seed germ-
ination, etc., many animal habitats are more or less
temporary, or at any rate tend to move about and
change their position. A burnt heather moor re-
generates new heather : if grazing is stopped it may
turn into a birch plantation, and later to a pine
wood, from which most of the previous stages of
succession have disappeared.

The subject of succession is dealt with by Leach
(1933) in another volume of this series. Its importance in animal ecology lies in two things : the
concept of shifting habitats which introduces a
dynamic *motif* into surveys of animal communities,
and the basis it gives for classifying vegetation
into a logical sequence of types related through
succession. Animal ecology itself is of great importance in the investigation of plant succession,
through such influences as grazing, destruction of
seed by rodents, and so on. It is partly for this
last reason that Phillips (1931) has stressed the idea
of *biotic communities*, since in the study of succession
it is often difficult to disentangle the effects of
plants and animals in causing changes in the soil
and vegetation.

We come now to the consideration of physical
factors. On land these can be divided roughly into
climatic and edaphic, the latter including the soil
and shelter, etc. In water, there are also chemical
factors to be considered. This phase of animal
ecology can be conveniently discussed from several
different angles. First, the study of physical habitats
themselves. Secondly, the co-ordination of such
observations with ecological surveys. Thirdly, the
manner in which these factors influence animals.
To these a fourth subject should be added, which
also brings in considerations of vegetation, and

(especially where parasites are concerned) the associated animals : how animals are enabled to find the type of habitat in which they are best suited to live. This problem leads on in turn to the question of <u>migration and dispersal</u>. It will be seen that these subdivisions of the subject of physical habitats form a natural sequence which would be followed in building up a complete explanation of the distribution of animals. Having accurately defined the habitats, we must describe the animal communities which live in them, and then seek an explanation of the action of the physical habitat factors in limiting distribution and numbers, while not forgetting that every animal species must possess some means of finding a suitable habitat for itself, in doing which dispersal is usually necessary.

The study of physical habitats has partly become split into separate sciences, as in the case of meteorology and to some extent of oceanography and soil surveys. Such sciences do not always give the data required in ecological work. Meteorology has aimed mainly at understanding the physical complex of the atmosphere. While it had obtained generalizations of value in ecology, such as the action of depressions as units of climate, and broad averages for the temperature and rainfall of various regions, it usually happens that the animal ecologist finds the methods of meteorology more valuable than the results. In addition, certain important techniques have been evolved by animal and plant ecologists to suit their work. Among these may be mentioned the Livingston atmometer for measurement of rates of evaporation, methods of obtaining the total radiant energy striking a black surface, for getting humidities in very small spaces such as holes in the ground and in walls, for measuring the temperature under bark and at various depths in tree trunks, etc. These and other methods are fully described by Shelford (1929), Chapman (1931), Buxton (1931, 1932), and others.

4

Buxton summarizes recent work on measurement and control of moisture. Corresponding methods have been developed for the study of aquatic conditions, and these have been summarized for fresh water by Chapman (1931). The science of limnology —the ecology of lakes—would require a volume to itself. It has received very great attention in Europe and in the United States. Work on this subject in Britain has now become centralized in the laboratory of the Freshwater Biological Association on Windermere.

We pass from this aspect of the problem, which largely concerns methods and contains only a few generalizations of importance in animal ecology, to the next point—the necessity of defining as accurately as possible the conditions under which animal communities actually live. A competent ecological survey attempts to record as fully as possible the general climatic data, geological and soil conditions, vegetation, with full lists of all plants, together with special notes on local variations. At the same time, it is necessary to run continuous observations by means of thermographs and other such instruments, to get an exact picture of the fluctuations in climate during day and night and at different times of year. The same applies to the vegetation, which will usually show important seasonal changes. We begin here to see that the habitat is not to be expressed in averages but in terms of the fluctuations that it displays, which will in turn be of great importance in showing limits, both in normal and abnormal years. The importance of such habitat records for all ecological surveys is that they enable comparisons to be made between similar communities in different regions, and between different communities in the same region. At the same time, such records soon acquire a historical value, as ecological succession and other factors bring about changes in the habitat, and with them changes in the animals. By the accumulation of such data, it is possible to build up circumstantial evidence as to the types of environment in which each species can live, and the important limiting factors controlling the numbers and distribution. For many species we have vague data, such as that they live in a dry habitat or are abundant in dry years and scarce in wet years. Ecological work tries to establish exact answers to the questions : at what season does dryness act, does it act directly, or indirectly through vegetation, or is dryness just an indicator of some other environmental factor such as temperature, how dry or how hot and for how long ? The circumstantial evidence obtained by correlation of animal communities or species with descriptions of their habitats is of great value in pointing the way, narrowing the field of inquiry for experimental work, which now has to be considered.

Most experimental work on physical factors has been concerned with methods and technique—still in an embryonic stage in many instances—and with one-factor experiments. This work, probably owing to the technical difficulties which have still to be solved, but also to other difficulties which will be mentioned, has as yet produced few theoretical generalizations of great value in ecology, although the practical uses of such work are often considerable. Earlier experimenters concerned themselves with discovering which factors in the environment were of importance in animal life, and with methods for their measurement. On land we have to measure and control for experimental purposes light (intensity, quality, length of day), heat (the temperature, and the total ' heat budget ' in a given time), moisture (the relative humidity, saturation deficiency, rainfall, etc.), other physical factors such as soil, substratum, shelter, etc. In fresh water and the sea there are parallel but often different methods of measuring light and heat, also methods for sampling gas contents, salts, hydrogen ion concentration, and so on. A large number of artificial experiments have been carried out in the laboratory. In these, all factors except one are controlled and the one factor varied. By such experiments it has been proved that every species of animal has very definite optimum conditions both for maximum survival (length of life), activity, sometimes reproductive rates, and population increase. Above and below the optimum there is a falling off in survival, activity, or rate of increase. Beyond certain points inactivity occurs and beyond that again death. The second method has been to combine two factors together, such as temperature and humidity. By doing this we obtain what is in effect a three-dimensional diagram, which can be plotted as a contoured graph. Thus in the tropical plague flea (*Xenopsylla cheopis*) the thermal death-point is 22° C. with relative humidity of 0 per cent. (completely dry), 27° C. with relative humidity of 30 per cent., 32° C. with relative humidity of 60 per cent., and 36° C. with a relative humidity of 90 per cent. (Mellanby, 1932). In other words the flea can stand higher temperatures in damper conditions. This shows the manner in which one physical factor conditions the action of another. Or take the marine flatworm (*Gunda ulvae*). This worm lives in the estuaries of streams running on to the sea-shore. It is subject alternately to fresh- and salt-water influence. Weil and Pantin (1931) showed that pure fresh water killed the worms in forty-eight hours, but that the onset of death (swelling up through intake of fresh water) could be prevented by adding a certain amount of calcium carbonate to the freshwater—as is the condition in nature.

A great many experiments of this type have been carried out, and the results so far achieved raise a hope that many features in animal distribution will be clearly explained in these terms. This kind of work is simply physiology applied to natural phenomena ; it is an attempt to explain in physiological (i.e. physico-chemical) terms the reactions of animals to ecological conditions. Attempts to obtain simple mathematical formulae for the temperature coefficients applicable to such experiments on animals have not so far been successful. The work of McLagan (1932) suggests one reason for this failure. He worked on the springtail (*Smynthurus viridis*) and found that different processes in the life of the springtail had different temperature and moisture optima. In other words, such an insect could find no environment to which it was completely adapted, and any attempt to express the temperature-humidity reactions in one formula would fail owing to heterogeneity of the processes abstracted.

Much of the laboratory technique has consisted of attempts to produce in the laboratory constant temperatures, constant humidities—in fact, artificially constant climates. It is, however, realized by most experimenters that these conditions do not accurately reflect the sort of climatic complexes found in nature. Climate is always fluctuating, and it remains to be discovered just how far the effects of a constant temperature or humidity on an animal approximate to those produced by a fluctuating environment with an average temperature or humidity of the same amount.

The *climograph* (invented by Ball in 1910 and named by Taylor in 1916) is a method of combining experimental data and field observations. It consists of a contoured diagram of the type already mentioned, expressing the combined effects of temperature and humidity (or other moisture index), and based upon the meteorological conditions occurring within the range of the animal studied. This diagram is then compared with the experimental data. The comparison makes it possible to find out whether an insect pest is already living within the full range of physical conditions possible for it. The method is of practical value in predicting either the future spread of an introduced pest, or the possible spread in favourable years when climatic limits have themselves expanded. Nichols (1933) worked out climographs for pure-bred flocks of British sheep and applied them tentatively to the forecasting of the regions in Australia and New Zealand most suitable for the establishment of these breeds. Climographs are a very convenient technique in the study of many problems, and they are fully discussed by Shelford (1929) and Chapman (1931).

In all this discussion it has been assumed that every animal is living in a certain habitat, and that physical and chemical conditions (as well as vegetation and associated animals) are those in which the animal can exist and breed successfully. But how does an animal find its habitat ? We enter here on a most intricate and interesting subject, which depends partly on a proper comprehension of the principles of animal psychology. It raises the whole question of the nature of animal reactions in general. Without going into psychological theories it may be said that there are four important questions which concern the animal ecologist. The first question is : Do animals have definite reactions which enable them to find the habitat suitable to their physiological and ecological characteristics ? Some animals are dispersed by broadcasting methods and arrive in large numbers in habitats both suitable and unsuitable to them. If they come to the right place they survive, if to the wrong place they die. Examples are young spiders spread by gossamer, and the floating larvae of many marine animals. Davis (1923) has shown that local variations in numbers of the mollusc (*Spisula subtruncata*) on the Dogger Bank are probably due sometimes to the spat falling on an unsuitable substratum. There remains, how-

ever, an enormous class of animals which have
special reactions for finding their normal habitat.
This finding may take the form of not going outside
the habitat they are in (avoiding the wrong one), or
of seeking new areas of the normal habitat by local
movements (e.g. butterflies laying eggs on certain
plants) or by long-distance migrations (e.g. woodcock
choosing certain woods for breeding in summer).
The second question is whether such actions are
adaptive ? By this is meant whether animals always
choose to live in the habitat to which they are best
suited, or whether they choose their habitat for
other reasons. This question is difficult to answer,
and twenty years ago, in the light of the evolution
theories of the day, would hardly have required an
answer. There is, however, a small but growing
body of evidence pointing to the existence of important elements of choice which depend as much upon
purely psychological factors as upon ecologically
adaptive reasons. Mention may be made of the
experiments of Thompson and Parker (1927) who
showed that certain polyphagous insect parasites do
not choose hosts according to any simple ecological
reasons that could be detected by the experiments,
which were very comprehensive. Lack (1933) has
brought forward important evidence that many birds
choose their habitats for reasons entirely unconnected with survival or breeding limits. He shows
quite convincingly that some Passerine birds adopt
certain habitats because they prefer them and for
no other reason. The gist of this work is that
animals probably do not usually inhabit all the
places which they could inhabit if they had different
ecological reactions. Against this it may be said
that we do not yet know enough about the limiting
factors of the lives of animals to make a negative
statement of this kind. The third question, one of
the greatest importance in practice, is whether such
reactions remain constant or whether they are

variable ? Again, there is a growing body of evidence that wild animals do not consistently frequent one habitat. Changes in habit are frequent, and we do not as yet know precisely what relative importance to attach to psychological factors (new ideas, or broken traditions or cumulative fatigue with old habits) and how much to organic changes in the form of mutations affecting behaviour. Finally, it is of great interest to inquire whether animals are actually conscious of their actions, and whether in this consciousness there is any element which is at variance with the usual concepts of animal behaviour current among physiologists and also many ecologists ? There is definite evidence that animals often migrate in response to stimuli which cannot be called danger signals but which appear to be simply unpleasant to them (Elton, 1930). Whether in this behaviour we can discern feelings akin to aesthetic feelings, or whether they are to be looked upon as mechanical upsets of mental balance, cannot be decided. The whole question of animal behaviour in relation to the choice of habitats and habits in general is of profound importance both in theoretical science and in practical economic biology. For instance, the herring shoals during their migration in the North Sea avoid certain patches of sea polluted by the colonial plankton flagellate (*Phaeocystis poucheti*), and the diatom (*Rhizosolenia sinensis*) (Savage, 1932). The variable position of these patches is an important factor in determining the success of East Anglian fisheries at certain times of year. The changing of its food plant by an insect may produce a new pest with far-reaching results. The kea of New Zealand changed from catching insects to scratching out the kidneys of sheep, and created an economic problem. A bird may change its reactions to habitat and by living in a new habitat alter the whole trend of evolution in its structure.

Much of the most significant modern work on the effects of physical conditions upon animals concerns the effects on numbers. To this problem of animal populations we now turn.

CHAPTER V

NUMBERS: STATISTICS

IT is convenient to consider problems of animal numbers under two headings. The first, discussed in the present chapter, covers our knowledge of the actual densities of numbers of various animals, the limits to increase, local groupings or aggregations of numbers, and methods of studying measuring and numbers in the field. This side of the population problem we have called *Statistics*. The other side of it is concerned with rates of increase, fluctuations in numbers, and the relation of problems of numbers to the environmental factors which influence the population. This we have called comprehensively *Dynamics* (Chapter VI). The statistics are what we find out about animal populations at any one moment. The dynamics introduce movements in time. The difference is something like that between the study of morphology and the study of development, between an instantaneous photograph and a moving picture. The periodic censuses that an animal ecologist carries out are to his science what the serial sections are to the embryologist or unit photographs to the cinematographer.

It is obvious, then, that before we can begin to draw any conclusions about numbers and the nature of animal population problems we require adequate census data to work with, just as ecological surveys produce the elementary data for understanding the organization of animal communities and the significance of animal inter-relations. Most of the urgent economic ecological problems are concerned with animal population problems : how many whales are left in the Antarctic seas, whether too many fish are being taken off the bed of the North Sea, whether the muskrat is spreading and increasing in England or Finland, whether a certain insect pest will be more or less abundant this year, how to check the increase of malaria organisms in the blood of infected human beings, whether the greatly increased trapping of snowshoe rabbits in Canada will affect the numbers of lynx which prey upon them, the causes of periodic decrease of partridges in England, and so on.

The stimulus to carry out censuses of wild animals has thus come to a great extent from the pressure of economic circumstances, often from general motives of stock-taking—the estimation of natural resources in a country. At the same time ecologists have been led on naturally from a study of natural animal communities towards a study of their population problems. These problems form the highest and most complex stratum of ecological work, and the science is yet in an early stage, through a lack of abundant facts with which to work. We may hope to see during the next two or three hundred years the development of a science of animal numbers which will take its place beside older sciences such as astronomy and chemistry that took equally long to reach a stage in which the power of prediction was established with certainty. This science of animal numbers is of very great importance in the task of combating various animal and plant pests which attack man and his domestic animals and plants, and wild species upon which he also depends for material resources or for amenities. At the same time a correct understanding and weighing of the evidence for the theory of natural selection and methods of evolution in general also awaits the further development of this phase of animal ecology.

Various methods of taking animal censuses have been evolved. They consist of complete counts or estimates of all the individuals in a given area. Where the animals are very small or very numerous or the area very large, complete sample counts are made at representative stations and used to calculate the total numbers on the larger area. The method of direct counts is used for some mammals, birds, and reptiles, and the method of sample counts usually for all smaller animals. Sampling has been applied with great success and accuracy to animal populations in the soil, in freshwater and marine plankton, also in bottom faunas of fresh water and the sea, and for the estimation of parasite densities. It also has to be used in a good deal of work on vertebrates, such as voles or small birds, or in the case of sea-bird colonies. Each of these branches of study is evolving suitable techniques which have so far been most highly developed in aquatic work and in soil investigations. Bird censuses have recently received much attention from naturalists and reference may be made to Nicholson (1932) who summarizes the methods that can be used. In England several large-scale bird censuses have been carried out by organizing the assistance of naturalists all over the country. Thus the British heron census (Nicholson, 1929) proved that there were about 8,000 breeding herons in England and Wales in the year 1928. A census of the great crested grebe (Harrisson and Hollom, 1932) showed that there were about 2,650 grebes in England and Wales. Nicholson (1932) has published maps summarizing the distribution of the density of these two aquatic birds in different parts of England. By such censuses it is possible to construct maps showing the contouring of density of numbers. The maps can then be correlated with various features in the environment which in turn suggest reasons for distribution and varying numbers. When such censuses are taken at periodic intervals it is possible to follow accurately the changes or fluctuations in the numbers of a species. A good deal of work has been done on the rook in this country. This has been summarized by Alexander (1933).

Another method of studying animal numbers is by making comparisons from month to month or year to year. If these comparisons can be made according to some standard method they are of considerable value in showing the *relative* numbers at different times. More will be said about this in the next chapter. For the moment let us inquire what conclusions of a general ecological importance have been obtained so far from the census work that has already been done on animals.

The results of censuses have to be expressed in terms of the number of animals on a given area at a given time, i.e. the *density of numbers*. But it is not sufficient to study only the crude densities. Suppose we are studying herons. The crude figures might give the number of herons on 50 square miles of country. This we may call the *Lowest Density* (Elton, 1932). It expresses the number of herons on 50 square miles of country including all kinds of habitats, some of which herons do not frequent at all. The Lowest Density expresses in a sense the biological success and adaptive elasticity of the species over a large area of mixed country. If there are a large number of ponds and lakes and streams there will be a chance for more herons to live in the area. If there are very few streams and ponds there will be very few herons. We therefore require another term which will allow for the patchy distribution of suitable habitats and express the numbers of herons in terms of the actual country that they really inhabit and breed in. This we call the *Economic Density*. The Economic Density will tend to be much more uniform in different parts of the heron's range than the Lowest Density. Within the natural habitat of the heron there are again local groupings or aggregations in density, brought about by chance crowding for abundant food supply or by social habits leading to communal roosting and nesting. It is convenient to use a third term—*Highest Density* —for these groupings, which are of importance ecologically since they play an important part in the spread of parasites and disease and in other ways. It often happens that these three ways of expressing density are convenient for different purposes. Thus the Lowest Density for partridges in an area of farming country is of great interest to the sportsman who wishes to know how many birds are available : the Economic Density gives the best measure of fluctuations in numbers from year to year ; while the Highest Density (e.g. the numbers in a covey roosting on a given patch of ground) is of importance in studying disease.

The definition of densities in these terms has given us a composite picture of the distribution of individuals in an animal population. We see that every species of animal tends to be distributed over any large area of varied country in two degrees or scales of density : the concentration into suitable general habitats, and the further local concentration into various forms of aggregation or of social groupings. The term *aggregation* is used in a general sense to cover local grouping of individuals. It may or may not be associated with actual social instincts or habits. Thus aggregations may be brought about by insects swarming to a suitable source of food, as bees and flies to fruit blossom, or by seeking common shelter, or surface for attachment, as with freshwater shrimps in weeds and under stones, or with barnacles on rocks. They may, on the other hand, be associated with regular instincts and habits, as when ants and bees form colonies, or rooks and herons nest together in groups. There is a very large and scattered literature on the subject of animal aggregations, and this has been brought together by Allee (1931), who has made two generalizations of great importance on the subject. Observations and experiments on a number of different animals have shown that they possess reactions which lead them to group together into aggregations, although the latter do not have any obvious significance or adaptive value socially, e.g. for defence or attack or the rearing of the young, and are not brought about simply by the mechanical influence of some common attractive factor such as food. Allee has carried out extensive studies, especially upon aggregations among freshwater isopods (*Asellus*). At the same time a great deal of other experimental work has been done on such widely different organisms as insects, isopods, echinoderms, protozoa, bacteria, yeasts, and also on spermatozoa. The results prove that slight crowding has marked effects in *increasing* the growth rates, survival rates or reproductive rates. Ecologists have become familiar with the effects of *overcrowding* in causing increased mortality from starvation or disease. These experiments on animal aggregations have proved beyond any doubt that *undercrowding* has also definite effects which are disadvantageous physiologically. For example, Pearl, Miner and Parker (1927) proved that the length of life of fruit-flies (*Drosophila*) was not greatest at the lowest densities but at densities higher than this. But still higher densities again caused lower average lengths of life.

These experimental studies by various workers agree in showing that there is usually an optimum population density for greatest physiological efficiency. There is still much discussion as to the reasons for this phenomenon in different instances. Among aquatic organisms discussion centres round the existence of various growth-promoting substances that may be produced by the organisms themselves, while other explanations have to be sought for terrestrial forms. We can see how there are also ecological reasons for very low densities being biologically inefficient for a species. For instance, if there were only one male and one female aphid in a ten-acre field the chance of their meeting and mating would be extremely small. The special interest of Allee's work lies in his suggestion that such physiologically conditioned aggregations (which he assumes to be ecologically advantageous also) give rise to conditions favourable to the evolution of more complex social organizations of the type discussed in Chapter III. The theory put forward is that such loose aggregations would tend to exploit through natural selection the potentialities of evolving into more powerfully knit groups, and finally into such dominant ' civilizations ' as those of the ants.

This concept of optimum densities can also be considered from another point of view. All animals tend to increase rapidly in numbers if their normal checks are temporarily removed. This sort of rapid increase is commonly observed during ' plagues ' of animals such as field-mice, cockchafers, locusts, and aphids. We have just seen that there are physiological reasons why the density tends to remain above a certain very low level. There are equally strong reasons preventing the density rising above a certain level. The ultimate limits to increase lie either in factors of space (as with some sessile marine intertidal animals such as the mussel *Mytilus* and the barnacle *Balanus*) or in the exhaustion of food supplies, as occasionally may be seen during plagues of caterpillars or in protected herds of deer in national parks. Thus, it has been found that one reindeer in Alaska requires about forty acres of pasture to maintain it continuously throughout the year. The phenomenon of over-parasitization sometimes occurs in insects, and leads to the destruction both of the host and the parasite through lack of food sufficient to complete development. Other limits which frequently come into force before space is filled or food used up, are the attacks of predators and parasites, the latter leading to outbreaks of disease, and to violent fluctuations in the population.

It is found that every species has a certain range of densities which may be called its optimum range. The range is greater than that found in human populations because animals are mostly shorter-lived and fluctuate more rapidly in response to disturbing factors in the environment. Usually a species does not remain at the optimum density but oscillates about this region. In using the term optimum we must be careful to distinguish between optimum *conditions* which may be physiologically or ecologically suitable for the species and the optimum *density* of its population. The optimum density is the highest density at which a species can remain without bringing into action automatic checks such as epidemics or starvation which will bring down its numbers to a level dangerously near extinction. For we are probably justified in assuming that a species which can maintain a consistently high average of numbers for a great many years has on the whole a better chance of evolving quickly than if its density is always very low. For, under conditions of ecologically stable high density, mutations must be more frequent; and also the chances of local extinction through external catastrophes are less than with low densities. In actual practice few species maintain constant populations at all, so that we do not usually find more than an approximation to the theoretically optimum range.

Probably owing to the interaction of physical adaptations and animal inter-relations (which do not always coincide in their incidence, i.e. are not always in the same habitat), many species do not live under the optimum conditions physiologically. Thompson (1929) has suggested that insects only become pests (i.e. very abundant) when all the conditions happen

5

to be at an optimum for increase in numbers. Of course such increase beyond a certain optimum density is not necessarily an adaptive advantage for the species, because it brings into action these automatic checks (such as epidemic disease) which may actually wipe out the species completely.

There is another way of studying densities of numbers. It is a general matter of observation that smaller animals make up by their numbers what they lack in size. Morris (1922) found the soil fauna of unmanured arable land to contain about 5,000,000 individuals per acre. Alexander (1932) has published figures for the numbers of birds per acre on farm land near Oxford. Examples are : song-thrush and linnet each 8 ; yellow-hammer and wood-pigeon each 17 ; blackbird, 35 ; and green woodpecker only 1. These are taken on a somewhat complex mosaic of habitats and therefore represent the Lowest Densities. Man has a much lower density than most of these birds. We can combine the facts about size and density by working out the *weights* of animals per acre. This can be done if we know the density of animals and the average weights of an individual. Thus a wood-ant (*Formica rufa*) weighs about 2 milligrams when dry. A nest-colony of 10,000 wood-ants (not an uncommon scale of size for a nest) would weigh about 20 grams. An acre of pine wood with 20 nests would have a total dry weight of ants of 400 grams—about a pound. Alexander (1932) converted his total bird census figures mentioned above into weights (not dry) per acre which came to averages of about 57 kilograms per acre in October, 47 in November, and 32 in February, the decrease representing in all probability the gradual winter mortality. Human density at one person to two acres would give a weight of about 6·5 kilograms per acre dry weight, and some 19·5 kilograms fresh. Morris (1922) estimated the total weight of nitrogen in the bodies of the soil fauna on unmanured land at Rothamsted to be about $7\frac{1}{2}$ lb. (3·4 kilograms) per acre. This method is valuable in enabling comparisons to be made between species of different sizes. It expresses to some extent the effective biological success, or at any rate the ecological importance of different species, and enables the relative importance of species occupying the same biological niche to be studied. There is at present little information along these lines and it is to be hoped that future work will supply facts about the total weights per acre of animals at different stages in a food-chain, or in different years.

We now see how ecological surveys and censuses could be combined to build up a picture of the total amount of animal tissue maintained on any given area, so that we should be able to calculate the actual biological turnover on different types of habitat and at different times of year. It might then be possible to express these ecological facts in terms of some physiological or biochemical unit such as the rate of oxidation or the rate of consumption of certain organic substances. Chapman (1931, 329) summarized some of the rather limited data which are available for estimating the total amounts of substance present in a whole animal community, counting all the species together. Three large American lakes (Mendota, Monona, and Waubesa) have plankton yields of 214, 238, and 215 lb. per acre of surface, or 1,974, 3,163, 4,398 milligrams of dry organic matter per cubic metre. This is the amount at one time. The annual total turnover of plankton in Lake Mendota came to about 10,700 lb. per acre (dry weight) or 107,000 lb. (about 48 tons) per acre of living plankton. The dry weights of benthic (bottom-living) animals at middle and deep zones in Lake Mendota were 43 lb. and 68 lb. per acre (p. 341). Another important generalization concerns the limits to summer growth of zooplankton, depending on limits to the amount of phytoplankton. The latter has been proved, both in the sea and in fresh water, to depend on the total amount of phosphorus available, which is exhausted at the end of the season.

CHAPTER VI

NUMBERS: DYNAMICS

'THE causes which check the natural tendency of each species to increase are most obscure. Look at the most vigorous species ; by as much as it swarms in numbers, by so much will it tend to increase still further. We know not exactly what the checks are even in a single instance. Nor will this surprise any one who reflects how ignorant we are on this head, even in regard to mankind, although so incomparably better known than any other animal. This subject of the checks to increase has been ably treated by several authors, and I hope in a future work to discuss it at considerable length, more especially in regard to the feral animals of South America.' These words were written by Charles Darwin about seventy-five years ago in *The Origin of Species*. The work on ecology of animals of which he speaks but which he never wrote, would have been of intense interest, since these problems of numbers lie at the very heart of all theories concerned with natural selection and the origin of new varieties and species. What progress has been made during these eighty-five years ? The next impetus to the study of the population problems of animals came from economic biologists. Hjort in 1904 was studying the fluctuations in numbers of the cod in European waters. From about 1870 onwards entomologists became involved in a long series of costly and difficult experiments in which parasites were used in order to

59

control introduced insect pests of agriculture and forestry. Medical entomology also added its quota of ideas about the inter-relations of animal numbers, as in the studies of Ross (1911) upon the malaria-carrying mosquito. More recently the transmission of diseases such as bubonic plague, tropical typhus, Rocky Mountain spotted fever, and tularaemia from mammals to man directly or by insect carriers, has focussed attention on the same kind of questions. A great deal of work upon economic problems is confined to special, almost watertight, compartments. Such subjects as plagues of field-mice, the carriage of plague by rats and other rodents, malaria and mosquitoes, the control of the European corn-borer moth in America or the sugar-cane leaf-hopper in Hawaii or gypsy moth in America, have huge literatures of their own. Ecologists are therefore faced with a scanty but fairly well-ordered literature on animal population studies by naturalists and ecologists, and by a vast and ill-co-ordinated and specialized literature of economic biology in which, however, many of the ideas and facts of use to the development of ecological theories may be found. Economic biologists are not uniformly conscious of the trends of animal ecology, but it should be stated at once that three of the most progressive contributions to the theories of animal numbers have arisen from economic investigations, in one instance on marine fisheries (Volterra,[1] 1926), and in the others on problems of insect pests (Thompson, 1924, etc., and Nicholson, 1933). Economic investigations will always supply a very large part of the facts from which ecological generalizations can be made, since their material is often collected on such a huge scale, as for instance in connexion with the North Sea fisheries, the experimental liberation of insect parasites, the international study of locusts, the routine examination of rats for plague at sea-ports, or the

[1] Based on the work of Ancona.

statistics available through the fur-trade. The earlier work of animal ecologists showed little realization of the importance of animal numbers, and animal ecology is now going through a stage in which it is being loaded with tremendous economic problems connected with the numbers of fishes and rodents and insects and other animals, while the purely scientific aspects of that subject are the least fully developed. We may therefore look forward to a period during which the studies of ecologists themselves will be greatly supplemented by material gathered for economic reasons and also by other material supplied by the labours of naturalists, who are in process of transferring to census work (collecting facts about the numbers of wild animals) much of the energy which they formerly applied (most usefully also) to the collection of specimens for study by systematists.

One of the most important generalizations that can be made about wild animal populations is that they fluctuate greatly in numbers. Naturalists of the nineteenth century took over without alteration the idea of the balance of life, i.e. constant populations. The earlier religious ideas had included the concept that the world was created in an orderly way, and disturbances in this order were attributed either to the acts of man himself or to the acts of God in punishing man for his presumption in upsetting this order, or perhaps in doing anything new at all. This general concept fitted naturally into the later biological theories of adaptation among animals, since it was supposed (rightly) that animals were closely adapted to their surroundings and (wrongly) that this adaptedness would lead to a state of steady balance between the numbers of different species. Whatever the various theories of adaptation may be found to require we are now in a position to show that wild animals do fluctuate very much and that these fluctuations are not simply the result of developments of human civilization in which man has interfered with natural conditions, as has happened with the introduction of pests into islands or the destruction of predatory animals. It is known, for instance, that animal life in Labrador is subject to violent periodic fluctuations of populations which affect a great many different kinds of animals—field-mice, foxes, ptarmigan, hawks, owls, caribou, and wolves, to name only a few. In this fluctuating nexus of animal communities man (Indians and Eskimos), far from being the interfering factor, is the unwilling victim of his environment, and in turn fluctuates in numbers through the ravages of starvation and disease (Elton, 1930, 1931b).

Some of these fluctuations are rather unexpectedly regular in their periodicity, others are irregular both in the period and the amplitude. Some examples may be given. Short-tailed field-mice or voles have periodic fluctuations in which disease plays an important part, both in Labrador, Great Britain, and Norway, and with a periodicity of about four years. Voles in Germany and France also fluctuate, but partly in connexion with agricultural conditions (Elton, 1931c). In the forests of northern Canada and also on the prairies of the Middle West, many of the smaller fur-bearing animals and rodents and game birds fluctuate in a rhythmical way, reaching abundance about every ten years, with slight variation in the exact periodicity. Johnstone (1928) showed that the marine fishes of the Mersey Estuary region between 1893 and 1927 (when records were available) fluctuated in a manner characteristic for each species. Thus plaice had maxima about 1895, 1910, and 1919 ; soles about 1898 and 1905 ; dabs about 1897, 1902–3, 1910, and 1924 ; and whiting about 1902, 1910, 1918, and 1925. Bottom-living marine animals of other kinds also suffer periodic fluctuations, as in the echinoderm (*Echinus miliaris*), which was destroyed in numbers on parts of the

British coast by cold winters in 1916–17 and 1928–9 (Orton and Lewis, 1931). Stenhouse (1928) found that house sparrows in the Shetlands mostly died from epidemic disease during the years 1926–8. The studies of Harrisson and Wynne Edwards (1932) on the island of Lundy show that the bird population of a limited area fluctuates greatly in different years, both in numbers and in species (i.e. some species become extinct, and others arrive). We are therefore left with a picture of animal communities as liable to many disturbing factors. Populations never remain constant for very long, and tend all the time to oscillate about a theoretical optimum point for the species. We shall have to inquire what happens when species depart very far from this theoretical optimum, what in general are the causes of these great fluctuations, and what effects they have upon associated animals—for we have already stressed the close interdependence of one form of animal upon another. When we reflect that most species are kept down in numbers in great measure by the attacks of enemies or parasitoids or parasites, we can see what profound influence the fluctuations in one animal must have upon others in the same community.

The actual recording of periodic changes in numbers can be carried on by two methods. In the first place, periodic censuses of the type described in the last chapter provide the most accurate measure of changes in *absolute* numbers of the population over a series of years. Such censuses have been carried out for rooks in certain parts of Great Britain (Alexander, 1933), and sampling censuses are widely used in plankton studies from year to year or in order to show seasonal changes and in the study of soil protozoa. Generally, it is not practicable to carry out complete censuses often enough or on a sufficiently large scale to provide absolute figures of density. Fluctuations can, however, be studied by another method : by recording the *relative* differences in numbers from season to season or year to year. This method is especially useful where a cyclical change of some regularity is being studied. The various techniques for following annual changes differ for each group of animals. For larger forms such as mammals and birds and fish, general subjective estimates of abundance are often a useful guide. The changes in numbers of the snowshoe rabbit (varying hare) in Canada during its ten-year cycle have been mapped by this means. Observers record their opinion as to whether rabbits are more or less abundant than in the previous year. Such opinions are of value when given by people who spend much of their life in contact with the animals concerned or make their living by hunting and trapping. A business man would remember whether prices were higher one year than another without necessarily remembering the exact figures. A trapper gets similar subjective estimates from his trapping experience. If a sufficiently large number of observers are taken, the method gives good results (Elton, 1933). The same methods can be adapted for observing game-birds (Gross, 1929–32), or fish, or crayfish (Duffield, 1933), or insects (Barnes, 1932). By this means the general nature of the periodicity of a species' fluctuations can be ascertained. The records provide a rough framework, a sort of reconnaissance of the problem, which is most valuable in directing more intensive research. And for a long time ecologists will have to employ these rough methods if only because observers of trained scientific experience are not numerous enough to obtain better results. Such observations do not always involve direct observation of animals, but may rely on traces, snow-trails, or other indications of abundance.

Rather more accurate measures of relative changes in numbers can be got by standard sampling of the population, where the sample taken is about the same percentage of the population each year, but the actual density is unknown. Trap-census work on mammals has given good results, e.g. with voles (Middleton, 1930 and 1931), while trapping has also been used for insects, e.g. blowflies that attack sheep. The flies per-boy-per-yard method employed in tsetse fly research is another example of an effective means of estimating relative abundance (cf. Nash, 1933 *a* and *b*). It is of value, not only in comparing changes in numbers from time to time, but densities in different habitats.

Fluctuations in numbers sometimes show both long and short components. For instance, the irruptions of Pallas's sand-grouse from the Gobi Desert into Europe (which are believed to be caused by periodic over-population) began apparently in 1863. They were repeated at intervals of about twenty-two years (major invasions), with smaller ones centring round the intermediate years. There is a suggestion that some factor started the periodic fluctuations after the middle of last century, after which they tended to fall into eleven- or twenty-two-year cycles (Elton, 1924 ; Thomson, 1926). The latest invasion was in 1909, and another is to be expected about the present time—unless the longer cyclical factor has ceased to operate and the invasions have ceased. Similar short-term fluctuations which arose and then died down again have been recorded for the cockchafer (*Melolontha*) in Denmark (Nusslin and Rhumbler, 1922), the grey squirrel in North America (Seton, 1920) and the springbuck in South Africa (Cronwright-Schreiner, 1925). Storrow (1932) has shown how a short-term fluctuation under the influence of a long-term trend of climate may gradually change its periodicity owing to a shifting of the optimum conditions for multiplication of the animal affected. We know that climate displays numerous pulsations, some regular or fairly regular, as with the two- to three-year pleionic cycle of tropical temperature studied by Arktowski (1916), or the thirty-five-year cycle in temperature and rainfall in European weather discovered by Brückner (1890).

A great many fluctuations are caused by factors primarily external to the life of the animal community. Examples of such factors are periodic frosts, hot or cold summers, destruction of food supplies through drought acting on vegetation, and so on. The simplest hypothesis to account for the widespread instability of animal populations is therefore that external disturbances in the weather, in vegetation, interference by man or by the arrival of new animals (whether by natural dispersal or the carriage of man) set up trains of internal disturbances, which upset the equilibrium of animal communities and make the populations of many species fluctuate. There is no doubt that this sort of thing does happen to a very great extent. The world contains broad differences of habitat, but does not supply constant environmental conditions, nor even regular cyclical conditions. The varied shape of the continents and oceans of the world, the variable amount of solar energy coming in as sunlight and heat, and the endless complicated results of these variations upon weather, the irregular mass-production of snow and ice which in turn produce further variations in weather, make a world far different from the thermostats of ecological laboratories.

There is, however, a very interesting and important recent group of theories, based in a large degree upon mathematical treatment of certain ecological assumptions. The main thesis of these theories is that the structure or organization of an animal community is such that it will in itself give rise to natural rhythmical fluctuations in numbers of the animals. The development of these theories is due to the work of Lotka (1920), Volterra (1926), Bailey (1931), and Nicholson (1933). They have approached the question by somewhat different methods. Thus, Volterra deals with cases of a predatory animal and its prey and makes certain assumptions about the dependence of reproductive rates upon the abundance of food. Nicholson, on the other hand, bases his theory upon considerations of the chances of predators (or parasitoids) searching for and finding their prey (or host) in different densities of population. Here again certain assumptions are made about the ecological reactions of animals while searching for food. Future work alone can decide whether the assumptions made in these mathematical studies are borne out by ecological investigations. There seems no reason to doubt that the purely mathematical side of the theories is sound. At present we can only say that there exists a very important possibility, suggested by these workers, that even if the physical conditions of the environment were completely constant, or at any rate varied with exact regularity, we should find in populations of wild animals oscillations that were independent of any regular oscillations in the environment. In fact, just as the human social and economic system has certain properties (depending upon the monetary system and the reactions and biological relationships of human beings organized in a certain complex but definite manner) that lead to the development of trade cycles or epidemics, so animal communities with their peculiar mode of organization so different from plants, but somewhat analogous to human communities, are subject to internal oscillations of population that result from and are set up from these inter-relations and the great powers of increase of animals.

It seems then likely, or at least possible, that fluctuations in numbers in animal communities may be brought about in two different ways : first by external disturbances in the environment, and secondly by internal oscillations peculiar to the community itself. But in both types of fluctuations the organization of animal inter-relations plays a dominant part. Nicholson has studied the probable interaction of the internal and external oscillations. This leads on to the second generalization that can be made about animal populations : the close inter-dependence of the numbers of one form upon those of another. No animal can fluctuate without affecting the numbers of some other animal. In many parts of Great Britain areas of pasture have been fenced off from sheep grazing and turned into young plantations. Within these fenced areas voles (*Microtus*) multiply greatly, and probably exist (in terms of total weight of animal) in quantities as great as the sheep were before enclosure. In 1891 and 1892 voles multiplied on grazing areas in the Border counties of Scotland, with the result that lambs starved and a Parliamentary Committee was set up to see what should be done about it (Maxwell and others, 1893). After the Committee had sat, the voles disappeared.

When lemmings multiply in Scandinavia or in the Arctic regions, there are usually great increases in predatory birds and animals such as skuas, snowy owls, hawks, ravens, foxes, ermines, etc. When the lemmings disappear, as they do after a year or two, these predatory animals starve, or do not breed for a year or two, or migrate elsewhere (Manniche, 1910). In tropical countries, multiplication of rats together with multiplication of rat-fleas (which depends on a combination of certain climatic conditions of temperature and humidity) causes an unusual multiplication of *Bacillus pestis*, and when the rats die, as well as before this, the fleas may migrate on to human beings and cause their population to fluctuate also, through bubonic-plague deaths.

We see, then, that the structure of an animal community, together with the variable environment that it lives in, leads both to fluctuations in individual species, and to far-reaching effects upon other species. It is frequently suggested that animal ecology can be divided into *autecology* (the study of single species in all their aspects) and *synecology* (the study of animal communities). It is clear that the study of the autecology of the numbers of any species involves inevitably a consideration of the synecology of the community in which it lives. It is for this reason that full ecological surveys are of such importance in ecology. It has already been shown how ecological surveys serve to define the animal inter-relations existing in a habitat. It is not true that a disturbance of the numbers of one species affects *all* the several hundred other members of the community in which it lives. It is true that we cannot easily predict that it will not indirectly affect some other species, apparently unconnected with it. We might easily find that some insect in town gardens was affected by the decrease in sparrows resulting from the decrease in horse-drawn vehicles and consequent scarcity of food. It has been suggested with some evidence that the decrease in summer diarrhoea in children in London may be due to the same cause—involving the decrease of horse manure in which breed house-flies that carry the germs of this disease (Graham Smith, 1930).

There is a fourth point of importance in connexion with the limiting factors to increase in animals. Factors which act as checks may be divided into two classes. Those which act irrespective of the density of the population. In this class come the effects of climate, of general food shortage not due to crowding, in fact, all external catastrophes. There is a second class of factors which depend upon the density of the population. An obvious example is disease due to a parasite and caused by overcrowding. And there are the factors (still mostly obscure) causing decreased viability or reproductive rates at very low densities (see Allee, 1931 and Chapter V). The distinction between these two modes of population control is very important. One acts in a random manner whether animals are abundant or scarce.

It kills the same percentage in either case. The other is related to the actual density of the species and usually acts as an automatic control. This idea of automatic control is important, since it explains to a large extent why animal communities continue to exist from year to year without striking changes in the composition of the species—in other words why species do not more often become locally extinct through chance fluctuations. There are various types of this automatic control. The increased chance of parasites finding or reaching a host as the density of the latter increases has been mentioned. Contrary to expectation, this does not always take place with parasites in mammals. Certain parasites have an age distribution in their hosts such that increase in numbers of the host causes temporary *decrease* in parasite density owing to the production of many young hosts which carry fewer parasites (Elton and others, 1931). In parasites such as coccidia that often attack young more than old, or both irrespectively, this damping down of automatic control by parasites does not occur (e.g. in the red grouse and the Norwegian willow grouse and the rabbit).

A very important factor in this general class of factors whose effect is to adjust the density of population of animals to their surroundings is *migration* (Elton, 1930). Migration is here used in the general sense of movements of individuals in the population. Heape (1931) has summarized with a valuable bibliography many of the known data on the subject of migration, and he makes a distinction between *migration* (periodic movements as in certain birds, marine and anadromous fish, and dragonflies), *emigration* (as in the Norwegian lemming when it leaves the mountains never to return), and *nomadism* (as in the caribou searching for good grazing grounds, or the bumble-bee searching for nectar-producing flowers). This distinction probably cannot be usefully upheld in the present state of our knowledge,

and the physiological explanations advanced by
Heape have as yet only limited evidence in their
support. The means by which animals find their
way about are also little understood, and information
on this point seems essential for classifying their
modes of migration.

In certain cases at any rate, no distinction can be
made between migration, emigration, and nomadism.
Thomson (1929) has shown that the woodcock in the
British Isles partly undertakes regular seasonal
migrations, partly emigrates in all directions, and
partly remains resident with local wandering movements. We are here concerned with the effect of
migration on the density of population. The drifting
movements of mixed flocks of small birds searching
for food in an English wood or South American forest
are good examples of the manner in which the
searching movements of animals smooth out local
differences in density of numbers. There is evidence
that in nearly all animals such searching movements
are of great importance in damping down the violence
of fluctuations in numbers. It is probable that without migration the continued existence of an animal
community would be impossible at its present degree
of complexity. Movements introduce a certain
elasticity in local densities of numbers, which is to
say that the psychological reactions of animals form
one of the chief problems of animal population.
Nicholson (1933) has worked out a theory containing
76 conclusions about numbers, based almost entirely
on considerations of the results of searching, i.e. of
migratory movements.

The spread of animals, in so far as it is not accomplished by passive dispersal, is due both to large-
scale emigrations, such as those of the lemming, and
to the accumulated effects of small local wanderings.
It is probable that some at least of the large-scale
migrations are due to real over-population, i.e. to
' pressure of numbers '. But the studies of Middleton

on the introduced American grey squirrel in England (1932) and of Harrisson and Hollom (1932) on the great crested grebe in England, have shown that spreading takes place equally in years of abundance and years of scarcity ; in other words, that a great deal of spreading may be due to accumulated local movements at the edge of the range, and not to pressure of numbers in the ordinary sense.

In the study of population problems two other aspects that have not so far been mentioned are of importance. The first is reproduction, and the second the length of life and age distribution of the population. Reproduction and the factors influencing it are mainly physiological problems, but they set the conditions for multiplication of any species. Ecologists have carried out important work on the influence of climatic and other factors on reproduction (see Chapman, 1931, and Uvarov, 1931), while the effects of undercrowding and overcrowding on animals have been summarized by Allee (1931). For mammals and birds there is a growing literature on the factors controlling breeding seasons, and the recent experiments of Rowan (1929) on birds and Baker (1932) on voles have shown the importance of the length of daylight and other less unexpected factors in controlling breeding. Baker (1929) has also studied the curious periodicity of breeding seasons in equatorial animals living under apparently constant conditions.

The second important variable—length of life—is of importance in fixing the replacement rate necessary to maintain or increase any population. It partly determines the rate of turnover of food stuffs and the living matter formed from them. Little accurate information exists about the length of life of wild species. Pearl and Doering (1933) have pointed out that such information is best charted in the form of a ' life curve ' of the kind used by insurance companies. This life curve shows the average number of individuals which survive to each age, and therefore also expresses the probable age distribution of the population at any moment. This age distribution, as has been mentioned, is of importance in controlling to some degree the density of parasites in a mammal population. Such life curves have been determined for one species of 'wild' animal in nature (man) and for at least two species of animals under laboratory conditions (the fly *Drosophila* and a rotifer *Proales*).

Our present knowledge of the dynamics of animal populations may thus be summed up as follows. The highest density for a species is not one which can long be maintained in nature, owing to the operation of automatic biological factors which bring about reduction in numbers, often far below that which could be permanently maintained. To a certain extent it is also true that very low densities cause a slowing down of increase in numbers. Reduction in numbers is also brought about very frequently by extraneous factors such as climatic variations, which cause catastrophes whose extent does not depend in any way upon the density of numbers at any particular time. Mathematical-biological theories have been developed to show that even in a constant physical environment fluctuations in numbers are to be expected in animal communities, these being caused by the inter-relations that exist between animals. If a single species of herbivorous animal free from enemies or parasites were to be kept in a constant environment, we should expect its population to increase either up to the exhaustion of food supply, or to supply its own intra-specific checks. The latter have been shown to exist in the flour weevil (*Tribolium confusum*) studied by Chapman (1928) and Holdaway (1932). Here the final level density is brought about by the fact that the weevils devour their own eggs if they chance to meet with them. A certain density of eggs and of weevils therefore results in the destruction of as many eggs as are being produced ; the final density varies with the conditions of each experiment (e.g. temperature). In an analogous manner the number of Arctic terns nesting on a certain island in Spitsbergen was limited by the reactions of the birds, which fight usually for a nesting territory of a particular size (though apparently the territory has little or no adaptive significance) (Summerhayes and Elton, 1928). In nature, animals live neither under constant physical conditions nor as isolated units. The result is that the populations fluctuate in a very complicated manner. The chief factors affecting the fluctuations are the type of inter-relations with other species, the optimum conditions physiologically for reproduction and existence, and the length of life. Migration is of great importance in smoothing out local differences in density and allowing of adjustment by the species to changing conditions.

Fluctuations in numbers have an important bearing upon evolution theories. Elton (1924, 1930) has suggested that they might allow non-adaptive (i.e. neither useful nor harmful) characters to spread in a population, and that it may be in this way that the apparently non-adaptive differences between closely allied species are established. Ford (1931) believes that no gene can exist in an animal without having some kind of physiological influence upon viability, and therefore that this explanation is not valid. Haldane (1932) has analysed the matter mathematically and concludes tentatively that genotypic variations may spread through fluctuations in the population, but at such a slow rate as to be negligible in the evolution of species. A partial bridge between these divergent points of view is the theory of Ford (1931) that useless or even normally non-viable characters may become established by spreading during the expansion phase of a population when competition is less severe, and may at the same time become modified by the interaction of genes and selection of a gene-complex which then alters the expression in characters of the originally harmful or useless genes. Ford's field studies (1930) of the butterfly (*Melitaea*) support his hypothesis. The mystery of the non-adaptive differences between closely allied species (that is to say the mystery of the origin of species) has still to be explained, but it is clear that population studies in the field and the laboratory have an important part in the further development of such evolution theories. A possible synthesis of the views already mentioned has been suggested by Elton (1931*a*). It is possible that many genes establish themselves (either in fluctuating or in relatively stable populations) through their adaptive importance to the viability of the organism during development. That is, a gene may be an important physiological component of a highly viable gene-family or gene-complex (see Ford, 1931), but still produce ecologically useless morphological characters. ' If a boy happened to be very ill while he was staying with a French family, the necessity for giving unselfish attention to him, or the money gained by looking after him, might make the family as a whole more successful morally or financially than they might otherwise have been. But this would not prevent the boy growing (perhaps as a result of the illness) into a little beast or merely very dull and useless ' (Elton, 1931*a*, 131–2).

Fluctuations in numbers are also of importance in causing alternately different types of natural selection, or at any rate varying their importance (Elton, 1924). The migration movements which are associated with population problems in animals also probably result in bringing the animals into contact with new environments and new situations, which may permanently modify their evolutionary trends (Elton, 1930). And finally, as was pointed out in the last chapter, the study of optimum population densities raises the whole question as to what is meant by biological success. Does it mean maintaining the highest density ?

An illuminating study by Nicholson (1927) shows the need for careful definition of terms when we speak of a new variation being *useful* to a species. It is normally assumed that such a useful variation will enable the species to increase in competition with other ones. Nicholson points out that if natural selection favours a cryptic (concealing) colour pattern protecting a butterfly from birds the immediate result will be the laying of eggs in higher density (since more butterflies escape attack). This results in a higher density of larvae, which in turn enables insect parasitoids to find more larvae in a given time and so produce a higher degree of parasitization. This finally brings down the density of adult butterflies which survive parasite attacks (these being often of the order of 90 per cent. of mortality causes). It is clear, therefore, that the original natural selection by birds may change the colour pattern, but does not necessarily increase the density of the population. Ecologically, nothing has been gained by the species except the possession of a new cryptic colour pattern. This process has been called ' intra-specific selection ' (Poulton, 1929). In actual fact, Nicholson believes that such inter-relationships will cause oscillations in numbers.

CHAPTER VII

ECONOMIC PROBLEMS

THE earliest annals of human history are full of the records of disease outbreaks in man or in his domestic animals, and also of ' plagues ' of animals such as field-mice, locusts, and beetles. These were specially well-marked manifestations of the fluctuations in population which we have seen to be almost universal among wild animals. In later years a new menace began to be added to this one. Originally the world had been split up by natural barriers into fairly well-limited zoogeographical areas. But the invention by man of better and better means of transport has had the unintended result of spreading round the world large numbers of animals whose arrival has often been the start of serious new pests or diseases. Massey (1933) has drawn attention to the new possibilities for the spread of yellow fever consequent on the development of air travel in tropical regions, especially in Africa. Yellow fever is a virus disease of man which attacks also certain monkeys, and is transmitted by the bite of mosquitoes which are specific hosts for the virus. Yellow fever used to be indigenous to West Africa, where it is comparatively mild. It spread to other regions, notably Central and South America. Although we have almost come to think of ' yellow jack ' as a vanishing disease, it is now realized that its spread into East Africa and from there to Asia has been in the past prevented mainly by transport barriers.

Ship transport was never fast enough for a person to reach a distant country while still in the stage at which mosquitoes can obtain the virus from the peripheral blood. Air travel makes this possible, and it is suggested that yellow fever might actually get to Asia, where the appropriate type of mosquito already exists naturally. This example illustrates the extreme dangers that modern transport engineering triumphs have introduced into the field of economic ecology.

Economic problems which come into the realm of ecology are to be found in thousands. They include the following main categories : *diseases* of man and of domestic animals ; *pests* of agriculture and stored products and forestry ; *fisheries*—including whale fisheries ; *conservation* of mammals and birds—including the fur trade and game research. These are the main categories, but there are others which have great importance in their own spheres. And, of course, the future of ecology as the accurate study of natural history and in relation to our general outlook upon wild animal life, is entitled to be considered as an important educational element.

Diseases of man have a special interest owing to the life histories of many parasites that have alternate hosts. *Bubonic plague* (sometimes spreading widely among human beings by direct breath infection as pneumonic plague) arises in all cases from rodents. These may be either ' domestic ' rats, or on the other hand may be marmots (Mongolia), susliks—a smaller marmot—(Russia), gerbilles (South Africa), ground squirrels (California), or guinea-pigs (Ecuador), or other forms (Wu Lien Teh and Pollitzer, 1926). The original contact between man and rodent is by fleas which carry the *Bacillus pestis*. *Tularaemia* is a fairly recently studied disease which also comes from rodents, either by contact with their dead bodies, or by the bites of insects such as Tabanid flies or by the bite of a tick. It occurs in North

America (hares, especially), the Volga region (water voles), Norway (hares), and other countries, but has not yet turned up in England. Plague kills most of those it attacks, and in the thirteenth century it killed more than half the population of Europe. Tularaemia (caused by a small bacterium) usually only kills less than one in twenty. Rats are now known to be the reservoirs also of *spirochaetal jaundice, rat-bite fever*, and certain forms of tropical *typhus*. *Japanese river fever* is a virus carried by a harvest mite (*Trombicula*) which has as chief host a vole, but also occurs on several other animals and on birds. Then *Rocky Mountain spotted fever* kills people by the bite of a tick that harbours the virus, the tick itself being dependent on wild rodents, etc., for its maintenance and bloodmeals (Parker, 1933). The great problem of Africa is *sleeping sickness* whose trypanosome is carried by Tabanid flies (*Glossina*) that bite man. The trypanosome can live also in wild game animals, while the fly can obtain blood from many wild animals such as game animals, also hippopotami, crocodiles, etc. Many parasites causing disease in man have only an insect as alternate host, e.g. *malaria* protozoa and *filaria* roundworms and *yellow fever* virus in mosquitoes, and *kala azar* protozoa and the *sandfly* fever virus in sandflies (*Phlebotomus*). *Schistosomiasis* is caused by a flat-worm whose alternate host is a snail, while the *guinea worm* is harboured in the larval stage by a Copepod, as also is the *tapeworm, Diphyllobothrium latum*. For a summary of part of this subject see Hull (1930).

It will be noticed that the alternate hosts are sometimes actually carriers of the disease organism, and sometimes only hosts of the insect or acarine vector. In all cases the numbers of each member of the complex are important in maintaining a complete cycle in which disease is possible. When we turn to domestic animals we find a similar story. Almost every human disease can be paralleled. Thus tsetse flies carry the trypanosome that causes *nagana* in cattle and horses in Africa, and this occurs also in game but does not harm them. *Red-water fever* is caused by a protozoan that is carried by ticks, which are also responsible in the tropics especially for the carriage of many other diseases. *Surra*, which kills camels by trypanosomiasis, is carried by flies. In South America *horse trypanosomiasis* is associated with periodic epidemics among capybaras—large water rodents (Joan, 1930). In Scotland, *louping ill*, a serious disease of sheep, is carried by a tick which also occurs on many wild animals and birds. The severe epidemics resembling *encephalitis* which occur among Arctic sledge dogs in the Arctic regions of Canada and Greenland are apparently associated with and sometimes derived from similar disease in wild Arctic foxes (Elton, 1931). *Gapes*, a round-worm disease of turkeys and fowls, is also harboured by starlings and rooks, and the latter frequently die of the disease in nature (Elton and Buckland). An-other *roundworm* is carried to horses by the agency of a horsefly. The *liver-fluke* of sheep is carried by a freshwater snail.

Turning to pests of agriculture the black list is found to be very long. The pests are partly indigenous ones such as the cabbage white caterpillars in England and partly introduced ones such as the colorado potato beetle in France. Locusts are one of the biggest problems in the tropics and the desert locust is said to have caused damage amounting to £6,000,000 during the last outbreak.

For a full account of agricultural pests the reader is referred to Wardle (1929), who gives useful summaries of the chief pests of each region of the world—including also notice of forestry and disease-carrying pests. The accounts given in this and other text-books leave an impression of world-wide struggle between growers of cotton, coffee, tea, fruits, vines, potatoes, corn and grain, and hundreds of other crops on the one hand, and thousands of insurgent insect pests on the other. Vast sums of money are lost through the destruction of crops, and very considerable sums are spent in attempts to combat the pests. To a certain extent such pest problems are cumulative : insects are constantly being carried round the world to new countries, in spite of stringent importation regulations and quarantine inspection. The impact also comes in from the spread of the cultivation of a particular kind of crop into regions where it was not known before. Apparently the olive fly (*Dacus oleae*) used to occur on wild olives in North Africa, and then spread into the cultivated olive plantations of southern Europe. On the other hand, the development of sugar-cane production on a large scale in Trinidad gave opportunities for an indigenous froghopper (*Tomaspis saccharina*) to become a pest by spreading the root fungi which cause Froghopper Blight. An enormous volume of economic biological research is produced annually, and may be followed in the *Review of Applied Entomology*, issued by the Imperial Institute of Entomology. The application of ecological ideas to these problems is as yet negligible, except perhaps in the realm of biological control and in the application of climograph methods.

Stored products in warehouses and factories and shops are also attacked by a great number of pests (Richards and Herford, 1930). Here the problem of control is different, owing to the possibilities of regulating the physical environment.

Forestry problems are always complicated much by the fear of periodic pest damage. Eidemann (1931) has shown that the incidence of defoliating tree-pests in the German forests during the last hundred years tended to fall into cycles of damage and freedom from damage. The working out of forestry operations on paper—and the profits form often a narrow margin and are calculated over a long period—is always complicated by this fear of destruction by pests, and by fire. Fruit orchards come on the border-line between forestry and agriculture, and it is here that some of the most spectacular pests have been studied and in certain cases controlled. Summaries of the regional problems connected with forestry and with fruit growing will be found in Wardle (1929) and in numerous textbooks of forestry.

We come now to a more positive side of economic ecology. The fisheries of the world were the first problem to be tackled seriously from the ecological standpoint, and it has taken some fifty to sixty years of co-operative work between West European countries to achieve a preliminary knowledge of the problem. The chief issue is the determination of policies in regard to the amounts that may be fished. It is estimated that at least a third of the fish population of the North Sea is taken every year. An idea of the intensity of fishing may be gained from the fact that most of the larger Pleistocene bone fossils on the surface of the Dogger Bank have been dredged up some time ago. A valuable summary of fishery ecological problems is given by Russell (1932). Similar problems arise in freshwater fisheries, in regard to overfishing, and also to the acclimatization of fishes in new countries. The salmon has special problems of its own, particularly in regard to pollution in the estuaries of large rivers, and to the periodicity in its numbers. Recently (after three hundred years of uncontrolled slaughter and destruction) the whale fisheries have been studied by modern ecological methods, during the *Discovery* expeditions. One of the indirect advantages of the world depression was the reprieve of a good many whales in the southern seas, owing to the fall in oil prices. The conservation of mammals and birds raises problems in connexion with national parks and sanctuaries and with sport. Two special economic aspects of conservation are sport (the production of good game crops and forecasting of bad ones) and the fur trade (in which both conservation of supplies and forecasting of fluctuations are important). Special attention to game problems has been given in Norway (the willow grouse), in New England and other parts of United States (grouse and quail), and in England (red grouse, partridge, etc.). The fur trade is a world-wide industry, but its fluctuations have been more especially studied in the Arctic regions and in the great conifer belt of the northern hemisphere (Canada, Alaska, Siberia, Russia, and Norway). Summaries of some of these investigations are given by Hewitt, 1921 ; Elton, 1924, 1931 ; Formosof, 1933 ; Johnsen, 1929.

To a certain extent these different economic problems interact. The conservation of game animals in Africa has had to reckon with the theories (probably incorrect) of medical investigators, who wish to wipe out game in order to destroy the reservoirs of nagana and sleeping sickness. White foxes are the most valuable furs in the Arctic, but their periodic disappearance is accompanied by epidemics which are apparently responsible for fatal disease in sledge dogs. Rats attack stored products and carry plague. Human diseases limit the areas in the tropics where white men can farm at all. River pollution affects both fisheries and water supplies. And so on.

We now have to consider the relation of ecological science to these problems. In the main we may say that the first important task of ecologists is to obtain *accurate information*. It frequently happens that practical policies of control of pests or fishing or methods of hygiene are based on fallacious preconceived ideas about the ecology of the animals. Thus the U.S. Fish Commission for many years bred marine fishes in aquaria and put them into the Atlantic Ocean to remedy a scarcity of fish which was part of a natural cycle in population. The measures were of negligible importance and merely masked the real reasons for natural recovery in fish such as cod. They were subsequently diverted with more success to inland fresh waters. During the great plague of London the authorities ordered all dogs and cats to be destroyed, as these were thought to be carriers, whereas they were in fact keeping down the real carriers—rats (Bell, 1924). Florence Nightingale ordered all windows in Indian hospitals to be thrown wide open, which would have produced remarkable results in a mosquito-ridden country that has over a million deaths from malaria per year. These examples are chosen from older historical accounts, but may of course still be matched anywhere at the present day.

With accurate information, practical policies can be formulated and nebulous theories can be given a proper perspective. An example may be given. In New Zealand the introduced trout were found to be getting smaller and smaller. The theories put forward by anglers were as follows : slower growth-rates caused by depletion of natural foods, which in turn were said to be due to silting of rivers through flooding due to cutting of forests, also due to introduction of foreign insectivorous birds which ate land insects so that they no longer fell into the rivers for trout to eat. There were other theories as well. Percival (1932) carried out a careful quantitative analysis of the situation, partly by measuring growth-rates and partly by ecological surveys of river faunas available as fish food. He found that all the theories were incorrect, and that there was a single reason for smaller trout, namely, that too many were being caught, and that they had not time or opportunity to grow into large ones.

In many realms of economic biology, however, the problem is not so much how to persuade authorities to undertake ecological investigations, as to

find effective means of applying the results of ecological discoveries in practical control. It is important to distinguish clearly between *economic ecological
problems* and the solution of these by *ecological
methods*. The first stage in attempted control of
the situation should always be the collection of
adequate information about the whole ecology of
the animals studied. This is the staff work. No
battle can be fought successfully without some
reasonably good information about the country and
the disposition of the enemy forces and his numbers
and the habits of his allies. Thus the habitats of
different tsetse flies are accurately known, something of their numbers in different habitats, their
life history, enemies, parasites, and the animals they
feed upon to get blood. The second stage is to
invent practical methods of control. This may or
may not be possible through ecological methods.
Thus an insect pest may be controlled in one case
by parasites (as with the gypsy moth), in others by
ploughing (as with the corn-borer moth), in others
by spraying with chemicals (as with many orchard
pests). The ecological information gathered by the
preliminary survey is necessary in order to direct
efforts on to the weak spot in the animal's life, but
this attack may be made either by ecological or by
other means. The tsetse fly has so far proved
resistant to biological control, although success is
being achieved by removal of its optimum habitat
by burning the bush.

   What are the ways in which animal ecology can
take part in the solution of these problems, apart
from general surveys for information purposes ?
One way is to follow the spread of introduced animals
and forecast (by climographs or other ecological
methods) the potential distribution. This has been
done by Cook (1924) for the pale western cutworm
in America ; by Nichols (1933) for pure-bred sheep
flocks (spreading under man's encouragement) in

England and New Zealand and Australia ; and by
McLagan (1932) for the springtail (*Smynthurus viridis*)
that attacks lucerne and alfalfa in Europe and
Australia.

The second thing is that natural fluctuations in
numbers can be studied and in some cases forecast.
Thus the ten-year cycle in fur-bearing animals in the
North Canadian forests can be forecast with some
accuracy, if a large enough region is taken (Elton,
1931*c*, 1933). This is also true of the shorter four-
year cycle in Arctic foxes in Canada (Elton, 1931*b*).
And the same methods have been developed for cockchafers in Europe (Zweigelt, 1928). Apart from know-
ing production or damage in advance, the recognition
of natural fluctuations is essential in order that the
effects of control measures can be estimated. Thus
the results of artificial introduction of mouse-typhoid
cultures in the control of field-mice in Germany and
France are vitiated unless the possibility of natural
epidemics in the mice is realized (Elton, 1931*c*). In
1930, introduced American grey squirrels died in
many parts of England (Middleton, 1931*b*, 1932).
At this time national efforts at destruction were
being planned. Had they been in operation for
several years the credit for the squirrel decrease
might have gone to them.

In actual control measures, only a few important
ecological ideas have as yet been exploited. Of
these the most interesting is *biological control*. Pests
that get to islands or to another continent often do
not have their natural parasites with them. By the
introduction of the appropriate parasites or similar
ones, control can be established. Huge sums of
money have been spent on this application of the
food-cycle organization of animal communities.
Sometimes the experiment has failed through the
existence of hyperparasites that limit the effective-
ness of the parasites. Good summaries of this work
are given by Imms (1931) and by Thompson (1930).

An analogous method is used for the control of pest plants, and has been used successfully for prickly pear and Lantana (Imms, 1931). The relation of insects to flowers is of great importance in horticulture and more especially in fruit orchards. In some cases the density of blossom is too great to allow of complete fertilization by the natural insect pollinator population, and bee-hives are planted at intervals to supplement the other insects.

When we come to conservation the practical problems turn to a large degree on fluctuations in numbers and on the maintenance of something approaching an optimum population. The aim of most ecological studies of game birds, wild mammals, etc., is to find some means of eliminating fluctuations and maintaining as high a density of numbers as is practicable without leading to disastrous reduction, e.g. through epidemics. The same aim is inherent in the fishery research. It is clear here that the concepts of animal numbers that are now emerging from ecological research have a most important bearing on these problems.

It will be clear that there is a difficulty in presenting in a small space a balanced account of economic ecology, since each of the subjects touched upon would fill an encyclopaedia. The situation may be summed up by saying that most economic biological problems deal with numbers of animals, and often with fluctuations in numbers and the inter-relations of different species of animals. Animal ecology is building up from basic surveys a science whose aim is the complete analysis of animal behaviour, numbers, and distribution. It has only recently progressed far enough to make close contact with the economic problems. It still has a long way to go.

REFERENCES

ADAMS, C. C. (1913). Guide to the Study of Animal Ecology. New York.

ALEXANDER, W. B. (1932a). *Trans. and Proc. Perthshire Soc. Nat. Sci.*, **9**, 35.

—— (1932b). *J. Animal Ecology*, **1**, 58.

—— (1933). *J. Animal Ecology*, **2**, 24.

ALLEE, W. C. (1923). *Biol. Bull.*, **44**, 167.

—— (1931). Animal Aggregations. Chicago.

ALVERDES, Fr. (1927). Social Life in the Animal World. London.

AMIRTHALINGAM, C. (1928). *J. Marine Biol. Ass.*, **15**, 605.

ARKTOWSKI, H. (1916). *Second Pan-American Sci. Congr.*, 1915–16, 1.

BAILEY, V. (1928). *Monogr. American Soc. Mammalogists*, No. 3.

BAILEY, V. A. (1931). *Q. J. Math.*, **2**, 68.

BAKER, J. R. (1929). Man and Animals in the New Hebrides, London.

—— (1932). *Proc. Royal Soc. London, B*, **110**, 313.

BALFOUR, A. (1922). *Parasitology*, **14**, 282.

BARNES, H. F. (1932). *J. Animal Ecology*, **1**, 191.

BELL, W. G. (1924). The Great Plague in London in 1665. London.

BIRD, R. D. (1930). *Ecology*, **11**, 356.

BOULENGER, E. G. (1929). *Proc. Zool. Soc. London*, 359.

BOYCOTT, A. E. (1921). *Proc. Malacol. Soc.*, **14**, 153.

BRISTOWE, W. S. (1931). *Proc. Zool. Soc. London*, 951 and 1383 ; *Ann. and Mag. Nat. Hist.*, **8**, 173.

BRÜCKNER, E. (1890). Klimaschwankungen seit 1700. *Geogr. Abhandlungen* (ed. by A. Penck), **4**, No. 2. Vienna.

BUCKLE, P. (1921). *Ann. Applied Biol.*, **8**, 135.

BUTCHER, R. W., PENTELOW, F. T. K., and WOODLEY, J. W. A. (1930). *Ministry of Agric. and Fish. Fishery Invest.*, Ser. 1, **3**, No. 3.

BUXTON, P. A. (1931). *Bull. Ent. Res.*, **22**, 431.

—— (1932). *Biol. Rev.*, **7**, 275.

CAMERON, A. E. (1913). *J. Econ. Biol.*, **8**, 159.
—— (1917). *Trans. Royal Soc. Edinburgh*, **52**, 37. London.
CARPENTER, G. D. HALE, and FORD, E. B. (1933). Mimicry. London.
CARPENTER, K. E. (1924). *Ann. Applied Biol.*, **11**, 1.
—— (1925). *Ann. Applied Biol.*, **12**, 1.
—— (1927). *J. Ecology*, **15**, 33.
CHAPMAN, R. N. (1928). *Ecology*, **9**, 111.
—— (1931). Animal Ecology with Especial Reference to Insects. New York and London.
COOK, W. C. (1924). *Ecology*, **5**, 60.
CRONWRIGHT-SCHREINER, S. C. (1925). The Migratory Springbucks of South Africa, 60. London.
DAKIN, W., and LATARCHE, M. (1913). *Proc. Roy. Irish Acad.*, B., **30**, 20.
DAVIS, F. M. (1923). *Ministry of Agric. and Fish. Fishery Invest.*, Ser. 2, **6**, No. 2 and (1925) **8**, No. 4.
DUFFIELD, J. E. (1933). *J. Animal Ecology*, **2**, 184.
EDWARDS, E. E. (1929). *Ann. Applied Biol.*, **16**, 299.
EIDMANN, H. (1931). *Ztschr. für Angewandte Entom.*, **18**, 537.
ELTON, C. (1924). *Brit. J. Exp. Biol.*, **2**, 119.
—— (1927). Animal Ecology. London.
—— (1930). Animal Ecology and Evolution. Oxford.
—— (1931*a*). *School Science Rev.*, 55 and 130.
—— (1931*b*). *Canad. J. Res.*, **5**, 673.
—— (1931*c*). *J. Hygiene*, **31**, 435.
—— (1932). *J. Animal Ecology*, **1**, 69.
—— (1933). *Canad. Field-Nat.*, **47**, 63 and 84.
ELTON, C., and BUCKLAND, F. (1928). *Parasitology*, **20**, 448.
ELTON, C., FORD, E. B., BAKER, J. R., and GARDNER A. D. (1931). *Proc. Zool. Soc. London*, 657.
FORD, E. (1923). *J. Marine Biol. Ass.*, **13**, 164.
FORD, E. B. (1931). Mendelism and Evolution. London.
FORD, H. D. and E. B. (1930). *Trans. Ent. Soc. London*, **78**, 345.
FORMOSOF, A. N. (1933). *J. Animal Ecology*, **2**, 70.
FOX, H. MUNRO (1932). *Nature*, **130**, 23.
FRASER, J. H. (1932). *J. Marine Biol. Ass.*, **18**, 69.
GRAHAM SMITH, G. S. (1930). *J. Hygiene*, **29**, 132.
GRINNELL, J., and STORER, T. I. (1924). Animal Life in the Yosemite. Berkeley, California.
GROSS, A. O. (1929–32). *Ann. Repts. New England Ruffed Grouse Invest.* Boston.
GURNEY, R. (1923). *J. Linn. Soc. London*, **35**, 411.
HALDANE, J. B. S. (1932). The Causes of Evolution. London.
HARDY, A. C. (1924). *Ministry of Agric. and Fish. Fishery Invest.*, Ser. 2, **7**, No. 3.

HARRISSON, T. H., and HOLLOM, P. A. D. (1932). *British Birds*, **26**, pp. 62, 102, 142, and 174.

HEAPE, W. (1931). Emigration, Migration and Nomadism. Cambridge.

HEWITT, C. G. (1921). The Conservation of the Wild Life of Canada. New York.

HINGSTON, R. W. G. (1932). A Naturalist in the Guiana Forest. London.

HOLDAWAY, F. G. (1932). *Ecol. Monogr.*, **2**, 261.

HOWARD, H. E. (1920). Territory in Bird Life. London.

HULL, T. G. (1930). Diseases Transmitted from Animals to Man. London.

IMMS, A. D. (1931). Recent Advances in Entomology. London.

—— (1931). Social Behaviour in Insects. London.

JOHNSEN, S. (1929). *Bergens Mus. Aarbok*, No. 2.

JOHNSON, M. S. (1926). *J. Mammalogy*, **7**, 245.

JOHNSTONE, J. (1928). *Proc. and Trans. L'pool Biol. Soc.*, **42**, 42.

JOAN, T. (1930). 5th *Reun. Soc. Argentina Pat. reg. Norte*, Jujay 1929, **2**, 1168. Buenos Aires (see *Rev. Appl. Ent.*, 1931, B, **19**, 7).

JORGENSEN, O. M. (1928). *Proc. Durham Univ. Phil. Soc.*, **8**, 41.

KASHKAROV, D., and KURBATOV, V. (1930). *Ecology*, **11**, 35.

KEEBLE, F. (1910). Plant Animals: a Study in Symbiosis. Cambridge.

KIRKPATRICK, R. (1917). *British Museum (Nat. Hist.), Econ. Ser.*, No. 7.

KROGERUS, R. (1932). *Acta Zool. Fennica*, No. 12.

LACK, D. (1933). *J. Animal Ecology*, **2**, 239.

LEACH, W. (1933). Plant Ecology. London.

LIGHT, S. S. (1926). *Ann. and Mag. Nat. Hist.*, **17**, 126.

LONGSTAFF, T. G. (1932). *J. Animal Ecology*, **1**, 119.

LOTKA, J. A. (1920). *Proc. Nat. Acad. Sci.*, **6**, 410.

McLAGAN, S. D. (1932). *Bull. Entom. Res.*, **23**, 101 and 151.

MANNICHE, A. L. V. (1910). *Meddel. om Grönlands*, 45, No. 1. Copenhagen.

MASSEY, A. (1933). Epidemiology in Relation to Air Travel. London.

MAXWELL, H. E., and others (1893). *Rept. Dept. Ctee. (Board of Agric.) on Plague of Voles in Scotland*, (C. 6943). London.

MELLANBY, K. (1932). *J. Exp. Biol.*, **9**, 222.

MIDDLETON (1930). *J. Ecology*, **18**, 156.

—— (1931a). *J. Ecology*, **19**, 190.

—— (1931b). The Grey Squirrel. London.

—— (1932). *J. Animal Ecology*, **1**, 166.

MORRIS, H. M. (1920). *Ann. Applied Biol.*, **7**, 141.

—— (1922). *Ann. Applied Biol.*, **9**, 282.

—— (1927). *Ann. Applied Biol.*, **14**, 442.

MURRAY, JAMES (1910). The Fauna of the Scottish Lochs. In *Bathymetrical Survey of the Scottish Lochs*, ed. by Sir J. Murray, Vol. 1. Edinburgh.

NASH, T. A. M. (1933a). *Bull. Ent. Res.*, **24**, 107.

—— (1933b). *J. Animal Ecology*, **2**, 197.

NICHOLS, J. E. (1933). *J. Animal Ecology*, **2**, 1.

NICHOLSON, A. J. (1927). *Australian Zoologist*, **5**, 10.

—— (1933). *J. Animal Ecology*, **2**, 132.

NICHOLSON, E. M. (1929). *British Birds*, **22**, 270 and 334.

—— (1931). The Art of Bird Watching. London.

—— (1932). *Discovery*, **13**, 309.

NUSSLIN, O., and RHUMBLER, L. (1922). Forstinsekentkunde, 342. Berlin.

ORTON, J. H., and LEWIS, H. M. (1931). *J. Marine Biol. Ass.*, **17**, 301.

PARK, O., LOCKETT, J. A., and MYERS, D. J. (1931). *Ecology*, **12**, 709.

PARKER, R. R. (1933). *Archives of Pathology*, **15**, 398.

PEARL, R., and DOERING, C. R. (1933). *Science*, **57**, 209.

PENTELOW, F. T. K. (1932). *J. Animal Ecology*, **1**, 101.

PERCIVAL, A. B. (1924). A Game Ranger's Notebook. London.

PERCIVAL, E. (1929). *J. Marine Biol. Ass.*, **16**, 81.

—— (1932). *New Zealand Marine Dept. Fisheries, Bull.* 5. Wellington, N.Z.

PERCIVAL, E., and WHITEHEAD, H. (1929). *J. Ecology*, **17**, 282.

—— —— (1930). *J. Ecology*, **18**, 286.

PETERS, B. G. (1930). *J. Helminth.*, **8**, 133.

PHILLIPS, J. (1931). *J. Ecology*, **19**, 1.

POULTON, E. B. (1929). *Proc. Zool. Soc. London*, 1928, 1037.

PRATT, E. (1898). *Ann. and Mag. Nat. Hist.*, **2**, 467.

RICHARDS, O. W. (1926). *J. Ecology*, **14**, 244.

—— (1927). *Biol. Rev.*, **2**, 298.

—— (1930). *J. Ecology*, **18**, 131.

RICHARDS, O. W., and HERFORD, G. V. B. (1930). *Ann. Applied Biol.*, **17**, 368.

ROCKHILL, W. W. (1894). Diary of a Journey through Mongolia and Tibet in 1891 and 1892, p. 9. Washington.

ROSS, R. (1910). The Prevention of Malaria. London.

ROWAN, W. (1929). *Proc. Boston Soc. Nat. Hist.*, **39**, 151.

RUSSELL, E. S. (1932). *J. Ecology*, **20**, 128.

SASSUCHIN and TIFLOV. (1932). *Vj. Microb. and Epid.*,
    11, 130 (summarized by Wu Lien Teh and R. Pollitzer
    (1932) in *Repts. National Quarantine Service*, Shanghai,
    Ser. 3, p. 200).
SAVAGE, R. E. (1926). *Ministry Agric. and Fish. Fishery
    Invest.*, Ser. 2, **9**, No. 1.
—— (1932). *J. Ecology*, **20**, 326.
SCOURFIELD, D. J. (1898). *Essex Naturalist*, **10**, 193.
—— (1908). *Intern. Rev. der Ges. Hydrobiol. u. Hydrogr.*
    1, 177.
SETON, E. T. (1920). *J. Mammalogy*, **1**, 53.
SHELFORD, V. E. (1913). Animal Communities in Temper-
    ate America. Chicago.
—— (1929). Laboratory and Field Ecology. Baltimore.
SOUTHERN, R., and GARDINER, A. C. (1926). *Ireland Fish.
    Sci. Invest.*, 1.
STENHOUSE, J. H. (1928). *Scottish Nat.*, p. 162.
STEPHEN, A. C. (1923). *Fishery Board for Scotland.
    Sci. Invest.*, 1922, No. 3.
—— (1933). *Trans. Royal Soc. Edinburgh*, **57**, 601.
STEPHENSON, T. A. and A., TANDY, G., and SPENDER, M.
    (1931). *Sci. Repts. Great Barrier Reef Expedition*,
    1928–9, **3**, No. 2.
STORROW, B. (1932). *J. Animal Ecology*, **1**, 160.
SUMMERHAYES, V. S., and ELTON, C. (1923). *J. Ecology*,
    11, 214.
—— (1928). **16**, 193.
SUMMERHAYES, V. S., COLE, L. W., and WILLIAMS, P. H.
    (1924). *J. Ecology*, **12**, 287.
TANSLEY, A. G., and ADAMSON, R. S. (1926). *J. Ecology*,
    14, 1.
TATTERSALL, W. M., and COWARD, T. A. (1914). *Mem. and
    Proc. Manchester Lit. and Phil. Soc.*, **58**, Nos. 8 and 9.
THOMPSON, M. (1924). *Ann. Applied. Biol.*, **11**, 349.
THOMPSON, W. R. (1924). *Ann. Fac. Sci. Marseille*, Ser. 2,
    2, 69.
—— (1929). *Parasitology*, **21**, 269.
—— (1930). *Empire Marketing Board (London), Publ.* 29.
THOMPSON, W. R., and PARKER, H. L. (1927). *Parasi-
    tology*, **19**, 1.
THOMSON, A. L. (1926). Problems of Bird-migration.
    London. 116–19.
—— (1929). *British Birds*, **23**, 74.
UVAROV, B. P. (1931). *Trans. Ent. Soc. London*, **79**, 1.
VOLTERRA, V. (1926). *Mem. Accad. Naz. Lincei (Sci. Fis.
    Mat. e Nat.)*, Ser. 6, **2**, No. 3.
—— (1931). Leçons sur la théorie mathématique de la
    lutte pour la vie. *Cahiers Scientifiques*, No. 7,
    Gauthier-Villars and Co., Paris. (See also Appendix
    by this author, in English, in Chapman, 1931.)

WALTON, C. L. (1913). *J. Marine Biol. Ass.*, **10**, 102.

—— (1925). *Ann. Applied Biol.*, **12**, 529.

WARDLE, R. A. (1929). The Problems of Applied Entomology. Manchester.

WARMING. (1896). Œcological Plant Geography. Berlin.

WEIL, E., and PANTIN, C. F. A. (1931). *J. Exp. Biol.*, **8**,
63 and 73.

WHEELER, W. M. (1922). Social Life among the Insects.
London.

—— (1928). The Social Insects. London.

WOODBURY, A. M. (1933). *Ecol. Monogr.*, **3**, 147.

WU LIEN TEH and POLLITZER, R. (1926). A Treatise on
Pneumonic Plague. Geneva.

WYNNE-EDWARDS, V. C., and HARRISSON, T. H. (1932).
J. Ecology, **20**, 371.

ZWEIGELT, F. (1928). Der Maikäfer. Suppl. to *Zeitschrift
für angewandten Entom.*, 13.

INDEX

Where the subject occupies more than one successive page, only the first is given.

Printed in Great Britain by
Jarrold & Sons, Ltd.
Norwich